REVIEW OF THE SCIENTIFIC APPROACHES USED DURING THE FBI'S INVESTIGATION OF THE 2001 ANTHRAX LETTERS

Committee on Review of the Scientific Approaches Used
During the FBI's Investigation of the 2001 *Bacillus anthracis* Mailings

Board on Life Sciences
Division on Earth and Life Studies

Committee on Science, Technology, and Law
Policy and Global Affairs Division

NATIONAL RESEARCH COUNCIL
OF THE NATIONAL ACADEMIES

THE NATIONAL ACADEMIES PRESS
Washington, D.C.
www.nap.edu

THE NATIONAL ACADEMIES PRESS   500 Fifth Street, N.W.   Washington, DC 20001

NOTICE: The project that is the subject of this report was approved by the Governing Board of the National Research Council, whose members are drawn from the councils of the National Academy of Sciences, the National Academy of Engineering, and the Institute of Medicine. The members of the committee responsible for the report were chosen for their special competences and with regard for appropriate balance.

This study was supported by Contract No. A9N0902700 between the National Academy of Sciences and the U.S. Federal Bureau of Investigation. Any opinions, findings, conclusions, or recommendations expressed in this publication are those of the authors and do not necessarily reflect the views of the organizations or agencies that provided support for the project.

International Standard Book Number-13:   978-0-309-18719-0
International Standard Book Number-10:   0-309-18719-2
Library of Congress Catalog Card Number   2011927648

Additional copies of this report are available from the National Academies Press, 500 Fifth Street, N.W., Lockbox 285, Washington, DC 20055; (800) 624-6242 or (202) 334-3313 (in the Washington metropolitan area); Internet: www.nap.edu.

Copyright 2011 by the National Academy of Sciences. All rights reserved.

Printed in the United States of America

THE NATIONAL ACADEMIES
Advisers to the Nation on Science, Engineering, and Medicine

The **National Academy of Sciences** is a private, nonprofit, self-perpetuating society of distinguished scholars engaged in scientific and engineering research, dedicated to the furtherance of science and technology and to their use for the general welfare. Upon the authority of the charter granted to it by the Congress in 1863, the Academy has a mandate that requires it to advise the federal government on scientific and technical matters. Dr. Ralph J. Cicerone is president of the National Academy of Sciences.

The **National Academy of Engineering** was established in 1964, under the charter of the National Academy of Sciences, as a parallel organization of outstanding engineers. It is autonomous in its administration and in the selection of its members, sharing with the National Academy of Sciences the responsibility for advising the federal government. The National Academy of Engineering also sponsors engineering programs aimed at meeting national needs, encourages education and research, and recognizes the superior achievements of engineers. Dr. Charles M. Vest is president of the National Academy of Engineering.

The **Institute of Medicine** was established in 1970 by the National Academy of Sciences to secure the services of eminent members of appropriate professions in the examination of policy matters pertaining to the health of the public. The Institute acts under the responsibility given to the National Academy of Sciences by its congressional charter to be an adviser to the federal government and, upon its own initiative, to identify issues of medical care, research, and education. Dr. Harvey V. Fineberg is president of the Institute of Medicine.

The **National Research Council** was organized by the National Academy of Sciences in 1916 to associate the broad community of science and technology with the Academy's purposes of furthering knowledge and advising the federal government. Functioning in accordance with general policies determined by the Academy, the Council has become the principal operating agency of both the National Academy of Sciences and the National Academy of Engineering in providing services to the government, the public, and the scientific and engineering communities. The Council is administered jointly by both Academies and the Institute of Medicine. Dr. Ralph J. Cicerone and Dr. Charles M. Vest are chair and vice chair, respectively, of the National Research Council.

www.national-academies.org

COMMITTEE ON REVIEW OF THE SCIENTIFIC APPROACHES USED DURING THE FBI'S INVESTIGATION OF THE 2001 *BACILLUS ANTHRACIS* MAILINGS

ALICE P. GAST (*Chair*), President, Lehigh University
DAVID A. RELMAN (*Vice Chair*), Thomas C. and Joan M. Merigan Professor, Stanford University School of Medicine, and Chief, Infectious Disease Section, Veterans Affairs Palo Alto Health Care System, CA
ARTURO CASADEVALL, Leo and Julia Forchheimer Professor of Microbiology and Immunology and Chair, Department of Microbiology and Immunology, Albert Einstein College of Medicine
NANCY D. CONNELL, Professor of Medicine, University of Medicine and Dentistry of New Jersey (UMDNJ)-New Jersey Medical School and Director, UMDNJ Center for BioDefense
THOMAS V. INGLESBY, Chief Executive Officer and Deputy Director of the Center for Biosecurity of University of Pittsburgh Medical Center, Associate Professor of Medicine and Public Health, University of Pittsburgh Schools of Medicine and Public Health
MURRAY V. JOHNSTON, Professor of Chemistry and Biochemistry, University of Delaware
KAREN KAFADAR, James H. Rudy Professor of Statistics and Physics, Indiana University
RICHARD E. LENSKI, John A. Hannah Distinguished Professor of Microbial Ecology, Michigan State University
RICHARD M. LOSICK, Maria Moors Cabot Professor of Biology, Harvard College Professor and Howard Hughes Medical Institute Professor
ALICE C. MIGNEREY, Professor of Chemistry and Biochemistry, University of Maryland, College Park
DAVID L. POPHAM, Professor of Microbiology, Virginia Polytechnic Institute and State University
JED S. RAKOFF, United States District Judge, Southern District of New York
ROBERT C. SHALER, Director, Forensic Science Program, Professor of Biochemistry and Molecular Biology, Pennsylvania State University
ELIZABETH A. THOMPSON, Professor of Statistics, University of Washington
KASTHURI VENKATESWARAN, Senior Research Scientist, California Institute of Technology Jet Propulsion Laboratory
DAVID R. WALT, Robinson Professor of Chemistry and Professor of Biomedical Engineering, Tufts University and Howard Hughes Medical Institute Professor

Staff

ANNE-MARIE MAZZA, Study Director
FRANCES E. SHARPLES, Study Co-Director (until 1/11)
ERICKA MCGOWAN, Program Officer, Board on Chemical Sciences and Technology (until 4/10)
STEVEN KENDALL, Associate Program Officer, Committee on Science, Technology, and Law
AMANDA P. CLINE, Senior Program Assistant, Board on Life Sciences (until 5/10)
KATHI E. HANNA, Consultant Writer
CAMERON H. FLETCHER, Editor

BOARD ON LIFE SCIENCES

KEITH R. YAMAMOTO (*Chair*), Executive Vice Dean, School of Medicine, and Professor, Department of Cellular and Molecular Pharmacology, University of California, San Francisco

BONNIE L. BASSLER, Investigator, Howard Hughes Medical Institute, and Professor of Molecular Biology, Department of Molecular Biology, Princeton University

VICKI L. CHANDLER, Chief Program Officer, Science, Gordon and Betty Moore Foundation

SEAN EDDY, Group Leader, Howard Hughes Medical Institute Janelia Farm Research Campus

MARK D. FITZSIMMONS, Associate Director, MacArthur Fellows Program, John D. and Catherine T. MacArthur Foundation

DAVID R. FRANZ, Vice President and Chief Biological Scientist, Midwest Research Institute

DONALD E. GANEM, Director, Global Infectious Disease Research, Novartis Institute of Biomedical Research

LOUIS J. GROSS, Professor of Ecology and Evolutionary Biology and Mathematics and Director, Institute for Environmental Modeling, Department of Ecology and Evolutionary Biology, University of Tennessee, Knoxville

JO HANDELSMAN, Howard Hughes Medical Institute Professor, Yale University

CATO T. LAURENCIN, Vice President for Health Affairs and Dean, University of Connecticut Health Center School of Medicine

BERNARD LO, Professor of Medicine and Director of the Program in Medical Ethics, University of California, San Francisco

ROBERT M. NEREM, Institute Professor and Parker H. Petit Professor Emeritus, Institute for Bioengineering and Bioscience, Georgia Institute of Technology

CAMILLE PARMESAN, Associate Professor of Integrative Biology, Section of Integrative Biology, University of Texas

MURIEL E. POSTON, Dean of Faculty, Skidmore College

ALISON G. POWER, Professor of Ecology and Evolutionary Biology and Dean, The Graduate School, Cornell University

BRUCE W. STILLMAN, President, Cold Spring Harbor Laboratory

CYNTHIA WOLBERGER, Professor, Department of Biophysics and Biophysical Chemistry, Johns Hopkins University School of Medicine

MARY WOOLLEY, President and CEO, Research!America

Staff

FRANCES E. SHARPLES, Director
JO L. HUSBANDS, Scholar/Senior Project Director
JAY B. LABOV, Senior Scientist/Program Director for Biology Education
KATHERINE W. BOWMAN, Senior Program Officer
MARILEE K. SHELTON-DAVENPORT, Senior Program Officer
INDIA HOOK-BARNARD, Program Officer
ANNA FARRAR, Financial Associate
CARL-GUSTAV ANDERSON, Program Associate
AMANDA MAZZAWI, Senior Program Assistant
AYESHA AHMED, Program Assistant

COMMITTEE ON SCIENCE, TECHNOLOGY, AND LAW

DAVID KORN (*Co-Chair*), Vice Provost for Research, Harvard University
RICHARD A. MESERVE (*Co-Chair*), President, Carnegie Institution for Science, and Senior Of Counsel, Covington & Burling LLP
FREDERICK R. ANDERSON, JR., Partner, McKenna, Long & Aldridge LLP
ARTHUR I. BIENENSTOCK, Special Assistant to the President for Federal Research Policy and Director, Wallenberg Research Link, Stanford University
BARBARA E. BIERER, Professor of Medicine, Harvard Medical School, and Senior Vice President, Research, Brigham and Women's Hospital
ELIZABETH H. BLACKBURN, Morris Herzstein Professor of Biology and Physiology, University of California, San Francisco
JOHN BURRIS, President, Burroughs Wellcome Fund
ARTURO CASADEVALL, Leo and Julia Forchheimer Professor of Microbiology and Immunology; Chair, Department of Biology and Immunology; and Professor of Medicine, Albert Einstein College of Medicine
JOE S. CECIL, Project Director, Program on Scientific and Technical Evidence, Division of Research, Federal Judicial Center
ROCHELLE COOPER DREYFUSS, Pauline Newman Professor of Law and Director, Engelberg Center on Innovation Law and Policy, New York University School of Law
DREW ENDY, Assistant Professor, Bioengineering, Stanford University, and President, The BioBricks Foundation
PAUL G. FALKOWSKI, Board of Governors Professor in Geological and Marine Science, Department of Earth and Planetary Science, Rutgers, The State University of New Jersey
MARCUS FELDMAN, Burnet C. and Mildred Wohlford Professor of Biological Sciences, Stanford University
ALICE P. GAST, President, Lehigh University
JASON GRUMET, President, Bipartisan Policy Center
GARY W. HART, Wirth Chair in Environmental and Community Development Policy, University of Colorado, Denver
BENJAMIN W. HEINEMAN, JR., Senior Fellow, Harvard Law School and Harvard Kennedy School of Government
D. BROCK HORNBY, Judge, U.S. District Court, District of Maine
ALAN B. MORRISON, Lerner Family Associate Dean for Public Interest and Public Service, George Washington University Law School
PRABHU PINGALI, Deputy Director of Agricultural Development, Global Development Program, Bill and Melinda Gates Foundation
HARRIET RABB, Vice President and General Counsel, Rockefeller University
BARBARA JACOBS ROTHSTEIN, Director, The Federal Judicial Center

JONATHAN M. SAMET, Professor and Flora L. Thornton Chair, Department of Preventive Medicine, Keck School of Medicine and Director, Institute for Global Health, University of Southern California
DAVID S.TATEL, Judge, U.S. Court of Appeals for the District of Columbia Circuit
SOPHIE VANDEBROEK, Chief Technology Officer and President, Xerox Innovation Group, Xerox Corporation

Staff

ANNE-MARIE MAZZA, Director
STEVEN KENDALL, Associate Program Officer

Preface

In autumn 2001, the tragic deaths, illnesses, and environmental contamination caused by the mailing of *Bacillus anthracis (B. anthracis)* spores in letters sent through the U.S. postal system caused tremendous fear and disruption in a nation shaken by the events of September 11. Efforts led by the Federal Bureau of Investigation (FBI) to characterize the material contained in the letters and identify the individual or individuals responsible for the mailings would involve extensive scientific study spanning almost nine years.

It is not unusual to use science to identify and characterize evidence and to link it to a particular individual in a criminal investigation. Indeed, in the 2001 *B. anthracis* mailings investigation, physics, chemistry, and biology all played a role. In this case, the field of bacterial genomics was rapidly evolving throughout the investigation. Recognizing the challenges inherent in such a complex scientific investigation, in 2008 the FBI asked the National Research Council (NRC) of the National Academy of Sciences (NAS) to conduct an independent review of the scientific approaches used during its investigation. In 2009, the committee, formed under the auspices of the National Academies' Board on Life Sciences and Committee on Science, Technology, and Law, began this review.

As we undertook this review, the committee kept in mind the context of the time, immediately following September 11, 2001, when there were multiple high-profile FBI investigations under way. We recognized that the grave consequences of these events for public health and national security and the uncertainty about possible additional attacks necessarily influenced the initial design and execution of the FBI's scientific investigation. Throughout its investigation, the FBI embraced and energetically pursued the use of new and emerging science through an unusual degree of involvement of outside scientists. In many ways, this case established and emphasized the potential importance of microbial forensics in the investigation of future acts of bioterrorism.

A scientific study is much more than a series of well-executed experiments. The planning and decision making used during a study are essential components of the science and can determine its outcome. As we learned from our review of this case, it is especially important in emergency situations to have a clearly defined approach for undertaking a complex scientific investigation involving many experts and collaborators. In any such case, the goal is to integrate a broad range of experimental methods and exploratory work with a more thorough and deeper study of selected methods, along with clear guidance for the investigation. Investigators must maintain a healthy degree of skepticism and a willingness to challenge their own assumptions. Achieving these goals requires considerable thought and planning before a crisis occurs. It is also important to recognize that when science meets law enforcement there are several tensions that need to be balanced: openness and secrecy, collaboration and independence, and deliberateness and expediency.

We also learned from this investigation that there is an immediate and ongoing need from the outset of an investigation to obtain expert advice and have available a group of advisors who can provide conceptual insight and relevant expertise to scientific plans, approaches, and scenarios.

An unavoidable observation from the 2001 *B. anthracis* mailings is that the best subject matter experts in a given area also might be viewed as suspects. Working with potential suspects during a sensitive investigation is a challenge that the law enforcement community must continually address through its vetting processes.

Throughout our review, we focused on the scientific aspects of the investigation and did not evaluate non-science-based investigative material. We have evaluated the science to the best of our ability, given the materials made available to us. While there may be additional relevant material to which we were not provided access, we believe that our review of the available material has resulted in many useful findings and conclusions. Nonetheless, other aspects of, and documents from, the FBI investigation may deserve future study and review.

In following our charge, we evaluated the specific conclusions drawn by the FBI based on its scientific analyses. The FBI never provided the committee with those conclusions in written form, although FBI conclusions were offered in oral presentations to the committee. We repeatedly sought written statements of conclusions until the case was closed by the Department of Justice (DOJ) on February 19, 2010. In our report, we address the conclusions offered in verbal reports, as well as the main scientific conclusions as written in the DOJ *Amerithrax Investigative Summary*.[1]

In November 2010, after our final report had been submitted to the FBI for a security review, the FBI informed the National Academies that there were additional materials relevant to the committee's work that had not previously

---

[1] United States Department of Justice. *Amerithrax Investigative Summary*. February 19, 2010. Available at: www.justice.gov/amerithrax/docs/amx-investigative-summary.pdf.

PREFACE   xiii

been shared with the committee. The Bureau offered to provide our committee with these materials and an additional briefing. After serious discussions with the National Academies' leadership, we agreed to receive and review these materials and reconvene the committee for one final meeting in January 2011. The documents and briefing provided us with additional information and led to meaningful changes in this report regarding the organization of the scientific investigation, sample collection, and analytical tests undertaken by the FBI and its contracting laboratories. This information resulted in the addition of a new section in the report (3.4.3) and the addition of a new finding (3.4) and recommendation (3.1). A benefit of the extension of the project and the delay in issuance of the report is that important additional materials, now available to the public, provide more information about the scientific investigation.

LOOKING TO THE FUTURE

While much of our effort was focused on a review of the science performed in support of the investigation of the 2001 *B. anthracis* mailings, an equally important aim has been to help ensure that future scientific investigations of biological attacks are conducted in the most relevant, rigorous, and thoughtful manner possible. Although the events of 2001 were tragic, they could have been more catastrophic. In the future, among many other requirements, it will be important to ensure more timely results, more efficient environmental analysis, access to globally representative strain collections, and a robust capability for characterizing less well studied or less easily cultivated biological agents. Officials also may need to manage expectations among the general public, policymakers, and the scientific community about the conclusions that can realistically be expected from the use of microbial forensics.

We have been fortunate to work with extremely talented, intelligent, and dedicated individuals in the undertaking of this multifaceted study. Committee members evaluated large numbers of documents under constrained circumstances that required exceptional dedication and patience. They listened intently to speakers, asked probing and insightful questions, and vigorously discussed what was learned, what we could research, and how to word our findings. We are indebted to them for all the time and energy they gave to this effort. We are also most grateful to the staff—Amanda Cline, Cameron Fletcher, Steven Kendall, Ericka McGowan, Anne-Marie Mazza, and Fran Sharples—and to the consultant writer, Kathi Hanna.

Alice P. Gast and David A. Relman
Chair and Vice Chair

Acknowledgments

This report has been reviewed in draft form by individuals chosen for their diverse perspectives and technical expertise, in accordance with procedures approved by the National Research Council's Report Review Committee. The purpose of this independent review is to provide candid and critical comments that will assist the institution in making its published report as sound as possible and to ensure that the report meets institutional standards for objectivity, evidence, and responsiveness to the study charge. The review comments and draft manuscript remain confidential to protect the integrity of the process. We thank the following individuals for their review of this report:

R. John Collier, Harvard Medical School
Rita R. Colwell, University of Maryland
M. Bonner Denton, University of Arizona
Ashlee M. Earl, Broad Institute of the Massachusetts Institute of Technology and Harvard University
Philip C. Hanna, University of Michigan Medical School
Stephen A. Johnston, Arizona State University
David H. Kaye, Arizona State University
Cato T. Laurencin, University of Connecticut Health Center
M. S. Meselson, Harvard University
Randall S. Murch, Virginia Polytechnic Institute and State University
Pauline Newman, U.S. Court of Appeals for the Federal Circuit
Stanley A. Plotkin, University of Pennsylvania (emeritus)
Elizabeth Rindskopf Parker, University of the Pacific
R. Paul Schaudies, GenArraytion, Inc.
James M. Tiedje, Michigan State University
Bruce Weir, University of Washington

Although the reviewers listed above provided many constructive comments and suggestions, they were not asked to endorse the conclusions or recommendations, nor did they see the final draft of the report before its release. The review of this report was overseen by Stephen Fienberg, Carnegie Mellon University, and Floyd Bloom, The Scripps Research Institute. Appointed by the National Research Council, they were responsible for making certain that an independent examination of this report was carried out in accordance with institutional procedures and that all review comments were carefully considered. Responsibility for the final content of this report rests entirely with the authoring committee and the institution.

Contents

SUMMARY   1

1   INTRODUCTION   25
    1.1   Background, 25
    1.2   Chronology of Events of Fall 2001, 26
    1.3   Brief Summary of the FBI's Scientific Investigation, 31
    1.4   Summary of FBI and DOJ Scientific Conclusions, 32
    1.5   Committee Process, 33
    1.6   Issues for Consideration in Reading This Report, 35
    1.7   Organization of the Report, 36

2   BIOLOGY AND HISTORY OF *BACILLUS ANTHRACIS*   37
    2.1   Introduction, 37
    2.2   The Biology of *B. anthracis*, 37
    2.3   Clinical Aspects of Anthrax, 38
    2.4   *B. anthracis* as a Biological Weapon, 40
    2.5   Phylogeny of *B. anthracis*, 41
    2.6   The Early History of the Ames Strain of *B.anthracis*, 44
    2.7   Summary, 44

3   SCIENTIFIC INVESTIGATION IN A LAW ENFORCEMENT CASE AND DESCRIPTION AND TIMELINE OF THE FBI SCIENTIFIC INVESTIGATION   47
    3.1   Introduction, 47
    3.2   Science and Scientific Investigation as Part of a Law Enforcement Investigation, 47
    3.3   The Federal Coordinated Response and Assignment of Laboratory Work, 55

3.4 Collection and Analysis of Clinical and Environmental Samples and Cross Contamination, 60
    3.4.1 Clinical and Epidemiological Samples, 60
    3.4.2 Crime Scene Environmental Samples, 64
    3.4.3 Samples from an Overseas Site Identified by Intelligence, 66
    3.4.4 Letter Material and Cross Contamination, 67
3.5 Committee Findings and Recommendations, 70

4 PHYSICAL AND CHEMICAL ANALYSES   75
    4.1 Introduction, 75
    4.2 Spore Preparation and Purification, 75
    4.3 Surrogate Preparation and Purification, 78
    4.4 Size and Granularity of the Material in the Letters, 79
    4.5 Presence of Silicon and Other Elements in the Letter Material, 80
        4.5.1 Elemental Analysis, 81
        4.5.2 Spatially Resolved Elemental Analysis, 83
        4.5.3 Silicon in the Spore Coat, 84
        4.5.4 Summary of the Silicon Analysis, 87
    4.6 Features of Bacterial Growth Conditions and Processing Methods: Detection of Meglumine and Diatrizoate, 87
    4.7 Media Component Analysis, 89
    4.8 Volatile Organic Compounds, 89
    4.9 Determining When the Material Was Produced: Radiocarbon Dating of *B. anthracis* Samples, 90
    4.10 Stable Isotope Analysis, 90
        4.10.1 *B. anthracis*, 90
        4.10.2 Water Samples, 92
        4.10.3 The Envelope Measurements, 92
    4.11 Committee Findings, 93

5 MICROBIOLOGICAL AND GENETIC ANALYSES OF MATERIAL IN THE LETTERS   97
    5.1 Introduction, 97
    5.2 Identification of the *B. anthracis* Strain, 97
    5.3 Was the *B. anthracis* in the Letters Genetically Engineered?, 100
    5.4 *B. subtilis* Contamination of the New York Samples, 104
    5.5 Identification and Characterization of Colony Morphological Variants in the Evidentiary Material, 106
        5.5.1 Why Was the FBI Interested in Colony Morphotypes?, 106
        5.5.2 Background Information on Morphotypes, 107
        5.5.3 Detection and Characterization of Morphotypes in the Anthrax Letters Samples, 109

            5.5.4   Selection Criteria for Genetic Variations Used in
                    Screening, 113
            5.5.5   Whole Genome Sequencing of Morphotype Isolates, 114
            5.5.6   Development and Application of Assays for the
                    Genotypes, 119
                        Genotypes A1 and A3, 119
                        Genotypes B and D, 119
                        Genotype E, 120
    5.6   Committee Findings, 121

6   COMPARISON OF THE MATERIAL IN THE LETTERS
    WITH SAMPLES IN THE FBI REPOSITORY   125
    6.1   Introduction, 125
    6.2   Creation of the FBI Repository (FBIR), 126
    6.3   Use of the Genetic Assays to Test for the Four Genotypes, 130
    6.4   Derivation of RMR-1029 Spores, 130
    6.5   Analyses of the Repository Samples and Statistical
          Interpretation of the Evidence, 132
            6.5.1   The FBI's *Statistical Analysis Report*, 135
            6.5.2   Committee Assessment of the *Statistical Analysis Report*, 136
    6.6   Analyses Based on Resampling RMR-1029 and Interpretation
          of the Results, 140
    6.7   Committee Findings, 144

BIBLIOGRAPHY   153

INDEX OF DOCUMENTS PROVIDED BY THE FEDERAL
BUREAU OF INVESTIGATION   161

APPENDIXES

A   Radiocarbon Dating   181
B   The Forensics Potential of Stable Isotope Analysis   183
C   Committee Evaluation of *Statistical Analysis Report*   185
D   Biographical Information of Committee and Staff   193

INDEX   205

Tables, Boxes, Figures

TABLES

S-1 FBI and DOJ Conclusions and Committee Comments, 11

1-1 Timeline of Key Events in the Anthrax Mailings Case, 28

3-1 Timeline of Scientific Events in the Anthrax Mailings Investigation, 48
3-2 Analytical Techniques Used on the Evidentiary Material, 58

4-1 Estimated Ranges of Total Number of Spores, 76
4-2 Estimates of Media Volume Required for Spore Preparation, 77
4-3 Methods for Chemical Analysis Referred to in Chapter 4, 81
4-4 Summary of Silicon Measurements in Evidentiary and Surrogate Samples, 82

5-1 Phenotypic Characteristics of the Morphotypes, 113
5-2 *B. anthracis* Isolates Analyzed by the Institute for Genomic Research (TIGR), 115
5-3 Further Genetic Characterization of the Morphotype Isolates, 116
5-4 Distribution Among the Anthrax Letters of the Genotypes Selected for Repository Screening, 118

6-1 General Results of the Screening of 1,059 Viable FBIR Samples for the Presence of the Mutation Genotypes, as Summarized by the Statistical Consultant to the FBI, 133
6-2 General Results of the Screening of the 947 Samples that Provided Definitive Results for All Four Genotypes, 134
6-3 Distribution Results for the Four Genotype Assays for Genotypes A1, A3, MRI-D, and E in the 947 Samples, 134

6-4 Observed and Expected (Under Independence) Distribution of Positive Signatures of Four Genotypes, 138
6-5 Genotype Assays on Three Replicates from Two Samples at 10 Dilution Levels: Entry Denotes Number of Positive Assays on Three Replicates at Each Dilution Level, 139
6-6 Results Obtained by Resampling from Flask RMR-1029, 143

C-1 Samples with Positive and "Inconclusive" or "Variant" Assays, 187
C-2 Probabilities of k 4-mutation Samples in Institution F, 191

BOXES

S-1 Charge to the Committee, 2

1-1 Charge to the Committee, 27

2-1 The Sverdlovsk Outbreak, 41

3-1 Bioterrorism Investigations, 54

5-1 Genome Sequencing, 101
5-2 The Polymerase Chain Reaction Technique, 102
5-3 The TaqMan Technique, 106

6-1 Subpoena Protocol for Collection and Submission of Ames Strain Samples, 127

FIGURES

2-1 Worldwide Distribution and Lineages of *B. anthracis*, 43

3-1 Trajectory and Outcomes of Anthrax Mailings, 62
3-2 *New York Post* Letter Powder, 63
3-3 Leahy Letter Powder, 63

4-1 SEM of Leahy and *New York Post* Powders, 85
4-2 Stable Isotope Results ^{18}O and ^{2}H, 91

5-1 *B. anthracis* Colony Morphotype "A", 110
5-2 *B. anthracis* Colony Morphotype "B", 111
5-3 *B. anthracis* Colony Morphotype "E", 112

A-1 Atmospheric CO_2 (Northern Hemisphere), 182

Summary

Less than a month after the September 11, 2001, attacks, letters containing spores of anthrax bacteria (*Bacillus anthracis*, or *B. anthracis*) were sent through the U.S. mail. Between October 4 and November 20, 2001, 22 individuals developed anthrax; 5 of the cases were fatal.

Although it was initiated as a public health investigation, the investigation quickly fell under the purview of the Federal Bureau of Investigation (FBI), when a deliberate act was suspected and letters containing *B. anthracis* were discovered in both New York and Washington, D.C., addressed to Tom Brokaw of NBC News, the *New York Post*, and U.S. Senators Tom Daschle of South Dakota and Patrick Leahy of Vermont.

Over the course of its investigation, the FBI devoted 600,000 investigator work hours to the case and assigned 17 Special Agents to a Task Force, along with 10 U.S. Postal Inspectors. The investigation spanned six continents; involved over 10,000 witness interviews, 80 searches, 26,000 email reviews, and analyses of 4 million megabytes of computer memory; and resulted in the issuance of 5,750 grand jury subpoenas. Additionally, 29 government, university, and commercial laboratories assisted in conducting the scientific analyses that were a central aspect of the investigation (U.S. Department of Justice [USDOJ], 2010, p. 4).

During its investigation of the anthrax mailings, the FBI worked with other federal agencies to coordinate and conduct scientific analyses of the anthrax letter spore powders, environmental samples, clinical samples, and samples collected from laboratories that might have been the source of the letter-associated spores. The agency relied on external experts, including some who had previously developed tests to differentiate among strains of *B. anthracis*.

In 2008, seven years into the investigation, the FBI asked the National Research Council (NRC) of the National Academy of Sciences (NAS) to conduct an independent review of the scientific approaches used during the investigation of the 2001 *B. anthracis* mailings (see Box S-1 for charge). In addition to informing FBI investigators about possible leads, much of the science used

BOX S-1
Charge to the Committee

The NRC was asked by the FBI to conduct an independent review of the scientific approaches used during the investigation. The official charge to the committee stated:

An ad hoc committee with relevant expertise will evaluate the scientific foundation for the specific techniques used by the FBI to determine whether these techniques met appropriate standards for scientific reliability and for use in forensic validation, and whether the FBI reached appropriate scientific conclusions from its use of these techniques. In instances where novel scientific methods were developed for purposes of the FBI investigation itself, the committee will pay particular attention to whether these methods were appropriately validated. The committee will review and assess scientific evidence (studies, results, analyses, reports) considered in connection with the 2001 *Bacillus anthracis* mailings. In assessing this body of information, the committee will limit its inquiry to the scientific approaches, methodologies, and analytical techniques used during the investigation of the 2001 *B. anthracis* mailings.

The areas of scientific evidence to be studied by the committee include, but may not be limited to:

1. genetic studies that led to the identification of potential sources of *B. anthracis* recovered from the letters;
2. analyses of four genetic mutations that were found in evidence and that are unique to a subset of Ames strain cultures collected during the investigation;
3. chemical and dating studies that examined how, where, and when the spores may have been grown and what, if any, additional treatments they were subjected to;
4. studies of the recovery of spores and bacterial DNA from samples collected and tested during the investigation; and
5. the role that cross contamination might have played in the evidence picture.

The committee will necessarily consider the facts and data surrounding the investigation of the 2001 *Bacillus anthracis* mailings, the reliability of the principles and methods used by the FBI, and whether the principles and methods were applied appropriately to the facts. The committee will not, however, undertake an assessment of the probative value of the scientific evidence in any specific component of the investigation, prosecution, or civil litigation and will offer no view on the guilt or innocence of any person(s) in connection with the 2001 *B. anthracis* mailings, or any other *B. anthracis* incidents.

in the investigation formed the basis of a rapidly developing but still nascent scientific field, called "microbial forensics," involving a series of laboratory tests used to determine the genetic identity of a microbial agent used for nefarious purposes. The development and application of microbial forensics would become an essential part of the scientific investigation in the hands of FBI investigators, who would combine it with physicochemical analyses to narrow their search for the possible origin of the anthrax used in the attacks. Key scientific questions focused on how, where, and when the material might have been produced; whether the material in all the evidence collected was identical; whether the material had been produced in such a manner as to make it more easily dispersible; whether it had any distinguishing physical or chemical properties of value in comparison studies; and whether its biological characteristics could provide leads to its origins.

The committee carried out its work mindful of the need to identify lessons that could be learned for future investigations in which science might play an important role.

Under the terms of the NRC contract with the FBI, the committee was initially provided with two boxes containing approximately 9,000 pages of materials regarding the scientific investigations undertaken by the FBI and by external experts contracted by the FBI during the investigation. At the end of the study an additional 641 pages were provided to the committee. Throughout the NRC study process, these FBI-provided materials were covered by FOIA Exemption 7, law enforcement, and were not publicly available. With the release of this report, as specified in the contract, these documents have been deposited in the NRC Public Access File.[1]

In addition to these materials the FBI briefed the committee on several occasions. Some of these briefings were done in open session, while others were conducted in closed sessions covered by FOIA Exemption 7. The committee also heard from a number of other experts.

In conducting its review, the committee was mindful that, while its focus was on the science involved in the case, the FBI did not ask the committee to review all of the science that was used in the investigation. For example, the committee was not charged to consider or evaluate any of the traditional forensic science methods and techniques used in criminal investigations (e.g., hair, fiber, fingerprint, or handwriting analysis) (NRC, 2009a) nor did it consider any of the psychological or behavioral sciences, such as linguistics, used by the FBI in its investigation. As such, this report and the committee's review and evaluation focused on the application of biological, physical, and chemical sciences to evidentiary materials from the letters, to the collection and analysis of environmental samples, to the analysis of the flask designated RMR-1029 (a flask containing

---

[1] The public can gain access to these materials by contacting the NRC Public Access Records Office.

anthrax spores that had been housed and maintained in a U.S. Army Medical Research Institute of Infectious Diseases [USAMRIID] laboratory at Fort Detrick since 1997), and to the collection and analysis of the *B. anthracis* samples from domestic and international laboratories and stored in the FBI Repository (FBIR).

During the course of the NRC committee's deliberations, the DOJ announced on February 19, 2010, that it was closing the case based on its conclusion that Dr. Bruce Ivins, a scientist at USAMRIID, had alone perpetrated the attacks. Dr. Ivins died on July 29, 2008, after taking an overdose of over-the-counter medications.

FBI SCIENTIFIC CONCLUSIONS AND COMMITTEE FINDINGS

The FBI drew a number of conclusions from its scientific investigation, which are summarized in Table S-1 at the end of the Summary. The committee found it challenging, however, to identify a definitive set of scientific conclusions drawn by the FBI investigators because they did not provide them in written form and because the conclusions provided publicly by DOJ in its briefings and Investigative Summary[2] varied from those provided by FBI officials in presentations to the committee. For the purposes of this report, the committee's analyses are based on the scientific conclusions provided by the FBI to the committee on September 24, 2009 (left-most column of Table S-1) and those issued publicly by DOJ on February 19, 2010, when it closed the case (USDOJ, 2010) (column second from the left in Table S-1). The committee was not in a position to offer a judgment about the importance and strength of the evidence from the scientific investigation relative to the importance and strength of the evidence from the criminal investigation because it was not charged with (and lacked the expertise for) reviewing the latter.

A summary of the committee's findings with regard to the scientific investigation and the scientific conclusions that were drawn from it by the FBI is provided below. The numbered statements below in bold that are labeled with an "S" (e.g., S.1) summarize subsets of the committee's findings and are intended to organize the findings and help guide the reader through them.

SUMMARY OF COMMITTEE FINDINGS

It is not possible to reach a definitive conclusion about the origins of the *B. anthracis* in the mailings based on the available scientific evidence alone.

S.1 The *B. anthracis* in the letters was the Ames strain and was not genetically engineered.

---

[2] United States Department of Justice. *Amerithrax Investigative Summary*. February 19, 2010. Available at: www.justice.gov/amerithrax/docs/amx-investigative-summary.pdf.

As background, the Ames strain of *B. anthracis* was originally isolated from a dead cow in Texas in 1981 and shipped to USAMRIID in Frederick, Maryland. Over time it was shared with research and development laboratories around the world.

- The dominant organism found in the letters was correctly and efficiently identified as the Ames strain of *B. anthracis*. The science performed on behalf of the FBI for the purpose of *Bacillus* species and *B. anthracis* strain identification was appropriate, properly executed, and reflected the contemporary state of the art. (Finding 5.1[3])
- The initial assessment of whether the *B. anthracis* Ames strain in the letters had undergone deliberate genetic engineering or modification was timely and appropriate, though necessarily incomplete. The genome sequences of the letter isolates that became available later in the investigation strongly supported the FBI's conclusion that the attack materials had not been genetically engineered. (Finding 5.2)

S.2 Multiple distinct colony morphological types, or morphotypes, of *B. anthracis* Ames were present in the letters. Molecular assays of specific genetic sequences associated with these morphotypes provided an approach to determining relationships among evidentiary samples.

As background, when bacteria are grown on agar plates, the descendants of single cells produce colonies. Variation in the appearance, or morphology, of these colonies (i.e., multiple morphotypes) can indicate the presence of different species or strains in the source material, each with a distinct genetic signature, or genotype.

- Multiple colony morphotypes of *B. anthracis* Ames were present in the material in each of the three letters that were examined (*New York Post*, Leahy, and Daschle), and each of the phenotypic morphotypes was found to represent one or more distinct genotypes. (Finding 5.4)
- Specific molecular assays were developed for some of the *B. anthracis* Ames genotypes (those designated A1, A3, D, and E) found in the letters. These assays provided a useful approach for assessing possible relationships among the populations of *B. anthracis* spores in the letters and in samples that were subsequently collected for the FBIR (see also Chapter 6). However, more could have been done to determine the performance characteristics of these assays. In addition, the assays did not measure the relative abundance of the variant morphotype mutations, which might have been valuable and could be important in future investigations. (Finding 5.5)

---

[3] The first number in the findings corresponds to the chapter in which they are presented.

- The development and validation of the variant morphotype mutation assays took a long time and slowed the investigation. The committee recognizes that the genomic science used to analyze the forensic markers identified in the colony morphotypes was a large-scale endeavor and required the application of emerging science and technology. Although the committee lauds and supports the effort dedicated to the development of well-validated assays and procedures, looking toward the future, these processes need to be more efficient. (Finding 5.6)
- A distinct *Bacillus* species, *B. subtilis*, was a minor constituent of the *New York Post* and Brokaw (New York) letters, and the strain found in these two letters was probably the same. *B. subtilis* was not present in the Daschle and Leahy letters. The FBI investigated this constituent of the New York letters and concluded, and the committee concurs, that the *B. subtilis* contaminant did not provide useful forensic information. While this contaminant did not provide useful forensic information in this case, the committee recognizes that such biological contaminants could prove to be of forensic value in future cases and should be investigated to their fullest. (Finding 5.3)

S.3 The FBI created a repository of Ames strain *B. anthracis* samples and performed experiments to determine relationships among the letter materials and the repository samples. The scientific link between the letter material and flask number RMR-1029 is not as conclusive as stated in the DOJ Investigative Summary.

- The FBI appropriately decided to establish a repository of samples of the Ames strain of *B. anthracis* then held in various laboratories around the world. The repository samples would be compared with the material found in the letters to determine whether they might be the source of the letter materials. However, for a variety of reasons, the repository was not optimal. For example, the instructions provided in the subpoena issued to laboratories for preparing samples (i.e., the "subpoena protocol") were not precise enough to ensure that the laboratories would follow a consistent procedure for producing samples that would be most suitable for later comparisons. Such problems with the repository required additional investigation and limit the strength of the conclusions that can be drawn from comparisons of these samples and the letter material. (Finding 6.1)
- The FBI faced a difficult challenge in assembling and annotating the repository of *B. anthracis* Ames samples collected for genetic analysis. (Finding 6.9)
- The results of the genetic analyses of the repository samples were consistent with the finding that the spores in the attack letters were derived from RMR-1029, but the analyses did not definitively demonstrate such a relationship. (Finding 6.2)

- Some of the mutations identified in the spores of the attack letters and detected in RMR-1029 might have arisen by parallel evolution rather than by derivation from RMR-1029. This possible explanation of genetic similarity between spores in the letters and in RMR-1029 was not rigorously explored during the course of the investigation, further complicating the interpretation of the apparent association between the *B. anthracis* genotypes discovered in the attack letters and those found in RMR-1029. (Finding 6.3)
- The flask designated RMR-1029 was not the immediate, most proximate source of the letter material. If the letter material did in fact derive from RMR-1029, then one or more separate growth steps, using seed material from RMR-1029 followed by purification, would have been necessary. Furthermore, the evidentiary material in the New York letters had physical properties that were distinct from those of the material in the Washington, D.C. letters. (Finding 4.6)
- The genetic evidence that a disputed sample submitted by the suspect came from a source other than RMR-1029 was weaker than stated in the Department of Justice *Amerithrax Investigative Summary*. (Finding 6.4)
- The scientific data generated by and on behalf of the FBI provided leads as to a possible source of the anthrax spores found in the attack letters, but these data alone did not rule out other sources. (Finding 6.5)
- Biological material from all four letters should have been examined to determine whether they each contained all four genetic markers used in screening the repository samples. (Finding 6.7)

S.4 Silicon was present in the letter powders but there was no evidence of intentional addition of silicon-based dispersants.

While any deliberate mailing of letters containing anthrax spores might be considered a form of spore "weaponization," this term has been more commonly used to describe preparations with enhanced properties of dispersion and aerosolization. It is commonly believed that deliberate efforts to make a powder more dispersible through the use of additives would suggest a more sophisticated level of preparation expertise. Thus, the presence of dispersants, such as nanoparticulate silica or bentonite, was an important feature in considering whether or not the letters contained "weaponized" anthrax spores.

- Although significant amounts of silicon were found in the powders from the *New York Post*, Daschle, and Leahy letters, no silicon was detected on the outside surface of spores where a dispersant would reside. Instead, significant amounts of silicon were detected within the spore coat of some samples. The bulk silicon content in the Leahy letter matched the silicon content per spore measured by different techniques. For the *New York Post* letter, however, there was a substantial difference between the amount of silicon measured in bulk

and that measured in individual spores. No compelling explanation for this difference was provided to the committee. (Finding 4.3)
- Surrogate preparations of *B.anthracis* did reproduce physical characteristics (purity, spore concentration, dispersibility) of the letter samples, but did not reproduce the large amount of silicon found in the coats of letter sample spores. (Finding 4.4)

S.5 It is difficult to draw conclusions about the amount of time needed to prepare the spore material or the skill set required of the perpetrator.

- The committee finds no scientific basis on which to accurately estimate the amount of time or the specific skill set needed to prepare the spore material contained in the letters. The time might vary from as little as 2 to 3 days to as much as several months. Given uncertainty about the methods used for preparation of the spore material, the committee could reach no significant conclusions regarding the skill set of the perpetrator. (Finding 4.1)

S.6 Physicochemical and radiological experiments were properly conducted to evaluate the samples for potential signatures connecting them to a source but proved to be of limited forensic value.

- The physicochemical methods used primarily by outside contractors early in the investigation were conducted properly. (Finding 4.2)
- Radiocarbon dating of the Leahy letter material indicates that it was produced after 1998. (Finding 4.5)

S.7 There was inconsistent evidence of *B. anthracis* Ames DNA in environmental samples that were collected from an overseas site. (Finding 3.4)

- At the end of this study, the committee was provided limited information for the first time about the analysis of environmental samples for *B. anthracis* Ames from an undisclosed overseas site at which a terrorist group's anthrax program was allegedly located. This site was investigated by the FBI and other federal partners as part of the anthrax letters investigation. The information indicates that there was inconsistent evidence of Ames strain DNA in some of these samples, but no culturable *B. anthracis*. The committee believes that the complete set of data and conclusions concerning these samples, including all relevant classified documents, deserves a more thorough scientific review.

S.8 There are other tools, methods, and approaches available today for a scientific investigation like this one.

- Investigators used reasonable approaches in the early phase of the investigation to collect clinical and environmental samples and to apply traditional microbiological methods to their analyses. Yet during subsequent years, the investigators did not fully exploit molecular methods to identify and characterize *B. anthracis* directly in crime scene environmental samples (without cultivation). Molecular methods offer greater sensitivity and breadth of microbial detection and more precise identification of microbial species and strains than do culture-based methods. (Finding 3.3)
- Point mutations should have been used in the screening of evidentiary samples. (Finding 6.6)
- New scientific tools, methods, and insight relevant to this investigation became available during its later years. An important example is high-throughput, "next-generation" DNA sequencing. The application of these tools, methods, and insight might clarify (strengthen or weaken) the inference of an association between RMR-1029 and the spores in the attack letters. Such approaches will be important for use in future cases. (Finding 6.8)
- The evidentiary material from this case is and will be immensely valuable, especially in the event of future work on either this case or other cases involving biological terrorism or warfare. It is critically important to continue to preserve all remaining evidentiary material and samples collected during the course of this (the anthrax letters investigation) and future investigations, including overseas environmental samples, for possible additional studies. (Finding 6.10)

S.9 Organizational structure and oversight are critical aspects of a scientific investigation. The FBI generated an organizational structure to accommodate the complexity of this case and received the advice of prominent experts.

- Over the course of the investigation, the FBI found and engaged highly qualified experts in some areas. It benefited from the unprecedented guidance of a high-level group of agency directors and leading scientists. The members of this group had top secret national security clearances, met regularly over several years in a secure facility, and dealt with classified materials. The NRC committee authoring this report, in keeping with a commitment to make this report available to the public, did not see these materials. (Finding 3.1)
- A clear organizational structure and process to oversee the entire scientific investigation was not in place in 2001. In 2003, the FBI created a new organizational unit (the Chemical, Biological, Radiological, and Nuclear [CBRN] Sciences Unit, sometimes referred to as the Chemical Biological Science Unit, or CBSU) devoted to the investigation of chemical, biological, radiological, and nuclear attacks. The formation of this new unit with clearer lines of authority is commendable. (Finding 3.2)

- As was done in the anthrax investigation, at the outset of any future investigation the responsible agencies will be aided by a scientific plan and decision tree that takes into account the breadth of available physical and chemical analytical methods. The plan will also need to allow for possible modification of existing methods and for the development and validation of new methods. (Finding 3.5)

LOOKING TO THE FUTURE

While much of the committee's effort was focused on a review of the science performed in support of the investigation of the 2001 *B. anthracis* mailings, an equally important goal has been to help ensure that future scientific investigations of biological attacks are conducted in the most relevant, rigorous, and thoughtful manner possible. Although the events of 2001 were tragic, they could have been more catastrophic. In the future, among many other requirements, it will be important to ensure more timely results, more efficient environmental analysis, access to globally representative strain collections, and a robust capability for characterizing less well studied or less easily cultivated biological agents. Officials may also need to manage expectations among the general public, policymakers, and the scientific community about the conclusions that can realistically be expected from the use of microbial forensics.

S.10 A review should be conducted of the classified materials that are relevant to the FBI's investigation of the 2001 *Bacillus anthracis* mailings, including all of the data and material pertaining to the overseas environmental sample collections. (Recommendation 3.1)

The committee did not receive nor review classified material. In November 2010 discussions with FBI and DOJ leadership regarding this report, we were made aware of additional information that would require review of classified material. Due to the lateness of this revelation, the importance placed on issuing a timely report, and the agreement between the NRC and the FBI that all materials we considered be publicly available, the committee did not undertake this additional review of classified material.

S.11 The goals of forensic science and realistic expectations and limitations regarding its use in the investigation of a biological attack must be communicated to the public and policymakers with as much clarity and detail as possible before, during, and after the investigation. (Recommendation 3.2)

TABLE S-1 FBI and DOJ Conclusions and Committee Comments

FBI conclusions	DOJ conclusions	Committee comment	Relevant report finding/section
	"Spores of such high concentration and purity indicate that they were derived from high quality spore preparations. Spores of this quality are often used in biodefense research, including vaccine development. It is important to have highly concentrated spores to challenge most effectively the vaccine being tested. Similarly, highly purified spores are necessary to prevent obstruction of the machinery used in those experiments. These findings meant that the anthrax mailer must have possessed significant technical skill" (USDOJ, 2010, p. 14).	The committee finds no scientific basis on which to accurately estimate the amount of time or the specific skill set needed to prepare the spore material contained in the letters. The time might vary from as little as 2 to 3 days to as much as several months. Given uncertainty about the methods used for preparation of the spore material, the committee could reach no significant conclusions regarding the skill set of the perpetrator.	Finding 4.1

continued

TABLE S-1 Continued

FBI conclusions	DOJ conclusions	Committee comment	Relevant report finding/ section
No substances were added to the spores after production to make them more dispersible (i.e., there was no "weaponization" of the spore material). Silicon and oxygen were present inside the spore coat and not on the outside (FBI, 2009).[a]	"Throughout the course of the investigation, repeated challenges have been raised to this finding that the spores were not weaponized. The challenges have their root in an initial finding by the Armed Forces Institute of Pathology ('AFIP') that, upon gross examination, the spores exhibited a silicon and oxygen signal. However, subsequent analysis of the spores by Sandia National Laboratories, using a more sensitive technology called transmission electron microscopy ('TEM') – which enabled material characterization experts to focus its probe of the spores to the nanometer scale – determined that the silica was localized to the spore coat within the exosporium, an area inside the spore. In other words, it was incorporated into the cell as a natural part of the cell formation process" (USDOJ, 2010, p. 14). Physical findings: Spore particles had a mass median diameter between 22 and 38 microns, exhibited an electrostatic charge, and were devoid of aerosolizing enhancers (e.g., fumed silica, bentonite, or other inert material) (USDOJ, 2010, p. 14).	Although significant amounts of silicon were found in the powders from the *New York Post*, Daschle, and Leahy letters, no silicon was detected on the outside surface of spores where a dispersant would reside. Instead, significant amounts of silicon were detected within the spore coats of some samples. The bulk silicon content in the Leahy letter matched the silicon content per spore measured by different techniques. For the *New York Post* letter, however, there was a substantial difference between the amount of silicon measured in bulk and that measured in individual spores. No compelling explanation for this difference was provided to the committee.	Finding 4.3

TABLE S-1 Continued

FBI conclusions	DOJ conclusions	Committee comment	Relevant report finding/ section
Isotopic analysis was conducted to determine the geographical source of evidentiary material based on water and grown media sources. Scientists at the University of Utah, under contract to the FBI, concluded that the results from the analysis were inconsistent with the spores having been produced at Dugway Proving Ground The FBI drew no conclusions based upon this analytical method (FBI, 2009). More generally, the results were inconclusive due to the large number of variables such as sources of water and the ratio of hydrogen and oxygen atoms found in culture medium, which made it difficult to narrow down an exact geographic source (FBI, 2009).		It was not possible to identify the location where the spores were prepared.	Finding 4.5

continued

TABLE S-1 Continued

FBI conclusions	DOJ conclusions	Committee comment	Relevant report finding/section
Tests led to the conclusion that two separate production batches of anthrax were used for the New York and Washington, D.C. mailings because each contained differences in spore concentrations, color, contaminants, texture, growth media remnants, and observed debris. When coupled with the genetic analysis discussed in Section B, *infra*, investigators were able to conclude that the two distinct batches of anthrax used in the 2001 attacks shared a common origin (FBI, 2009).	"Two separate production batches of anthrax were used for the New York and Washington, D.C., mailings because each contained differences in spore concentrations, color, contaminants, texture, growth media remnants, and observed debris. [But] when coupled with the genetic analysis, investigators were able to conclude that the two distinct batches of anthrax used in the 2001 attacks shared a common origin" (USDOJ, 2010, p. 15).	The flask designated RMR-1029 was not the immediate, most proximate source of the letter material. If the letter material did in fact derive from RMR-1029, then one or more separate growth steps, using seed material from RMR-1029 followed by purification, would have been necessary. Furthermore, the evidentiary material in the New York letters had physical properties that were distinct from those of the material in the Washington, D.C. letters.	Finding 4.6
	"This strain, known as 'Ames,' was isolated in Texas in 1981, and then shipped to USAMRIID, where it was maintained thereafter."(USDOJ, 2010, p. 3).	The dominant organism found in the letters was correctly and efficiently identified as the Ames strain of *B. anthracis*.	Finding 5.1

TABLE S-1 Continued

FBI conclusions	DOJ conclusions	Committee comment	Relevant report finding/section
There was no deliberate genetic engineering of the *B. anthracis* strain (FBI, 2009).	"The spore particles… showed no signs of genetic engineering" (USDOJ, 2010, p. 14).	The initial assessment of whether the *B. anthracis* Ames strain in the letters had undergone deliberate genetic engineering or modification was timely and appropriate, though necessarily incomplete. The genome sequences of the letter isolates that became available later in the investigation strongly supported the FBI's conclusion that the attack materials had not been genetically engineered.	Finding 5.2

continued

TABLE S-1 Continued

FBI conclusions	DOJ conclusions	Committee comment	Relevant report finding/section
The *B. subtilis* contamination found in the New York samples "did not provide...a productive avenue in terms of...leads for the investigation" (FBI, 2009).		A distinct *Bacillus species, B. subtilis,* was a minor constituent of the *New York Post* and Brokaw (New York) letters, and the strain found in these two letters was probably the same. *B. subtilis* was not present in the Daschle and Leahy letters. The FBI investigated this constituent of the New York letters and concluded, and the committee concurs, that the *B. subtilis* contaminant did not provide useful forensic information. While this contaminant did not provide useful forensic information in this case, the committee recognizes that such biological contaminants could prove to be of forensic value in future cases and should be investigated to their fullest.	Finding 5.3
There were mainly wild-type *B. anthracis* Ames strain, but there were significant numbers of phenotypic variants or substrains (FBI, 2009).		Multiple colony morphotypes of *B. anthracis* Ames were present in the material in each of the three letters that were examined (*New York Post*, Leahy, and Daschle), and each of the phenotypic morphotypes was found to represent one or more distinct genotypes.	Finding 5.4

TABLE S-1 Continued

FBI conclusions	DOJ conclusions	Committee comment	Relevant report finding/ section
These phenotypic variants could be detected by a combination of assays for four different insertion/deletion polymorphisms (FBI, 2009).	"Genetic analysis of morphological variants identified mutations which were later exploited to develop specific assays to identify the presence of identical mutations in evidence collected during the investigation" (USDOJ, 2010, pp. 24-25).	Specific molecular assays were developed for some of the *B. anthracis* Ames genotypes (those designated A1, A3, D, and E) found in the letters. These assays provided a useful approach for assessing possible relationships among the populations of *B. anthracis* spores in the letters and in samples that were subsequently collected for the FBI Repository (FBIR) (see Chapter 6). However, more could have been done to determine the performance characteristics of these assays. In addition, the assays did not measure the relative abundance of the variant morphotype mutations, which might have been valuable and could be important in future investigations.	Finding 5.5
		There was inconsistent evidence of *B. anthracis* Ames DNA in environmental samples that were collected from an overseas site.	Finding 3.4

continued

TABLE S-1 Continued

FBI conclusions	DOJ conclusions	Committee comment	Relevant report finding/ section
	"…fifteen domestic laboratories and three foreign laboratories were identified as repositories of Ames strain anthrax at the time of the letter attacks." (USDOJ, 2010, p. 17). "…the collection of Ames isolates from laboratories both from the United States and abroad that constitute the FBIR are a comprehensive representation of the Ames strain (USDOJ, 2010, p. 28). "A total of 1, 070 samples were ultimately submitted [to the FBIR], which represents a sample from every Ames culture at every laboratory identified by the FBI as having Ames strain" (USDOJ, 2010, p. 24).	The FBI repository was developed from an intensive effort to identify laboratories having access to the Ames strain; however, we cannot conclude that this approach identified every laboratory or was a comprehensive representation.	Section 6.2

TABLE S-1 Continued

FBI conclusions	DOJ conclusions	Committee comment	Relevant report finding/section
	"FBI collaborated with various experts…to provide a clear and thorough protocol for the preparation of the repository submissions" (USDOJ, 2010, p. 77).	The FBI appropriately decided to establish a repository of samples of the Ames strain of *B. anthracis* then held in various laboratories around the world. The repository samples would be compared with the material found in the letters to determine whether they might be the source of the letter materials. However, for a variety of reasons, the repository was not optimal. For example, the instructions provided in the subpoena issued to laboratories for preparing samples (i.e., the "subpoena protocol") were not precise enough to ensure that the laboratories would follow a consistent procedure for producing samples that would be most suitable for later comparisons. Such problems with the repository required additional investigation and limit the strength of the conclusions that can be drawn from comparisons of these samples and the letter material.	Finding 6.1

continued

TABLE S-1 Continued

FBI conclusions	DOJ conclusions	Committee comment	Relevant report finding/ section
	"In 2007, after several years of scientific developmens and advanced genetic testing coordinated by the FBI Laboratory, the Task Force determined that the spores in the letters were derived from a single spore-batch of Ames strain anthrax called "RMR-1029" (USDOJ, 2010, p. 5.) "Later in the investigation, when scientific breakthroughs led investigators to conclude that RMR-1029 was the parent material to the anthrax powder used in the mailings,..." (USDOJ, 2010, p. 6). "After a time-consuming process, the scientific analysis coordinated by the FBI Laboratory determined that RMR-1029, a spore-batch created and maintained at USAMRIID by Dr. Ivins, was the parent material for the anthrax used in the mailings." (USDOJ, 2010, p. 8). ...genetic analysis led to the conclusion that RMR-1029, located at USAMRIID, was the parent material mailed so the mailed spores,..: (USDOJ, 2010, p. 16 "...based on advanced genetic testing combined with rigorous investigation, the FBI concluded that RMR-1029 is the parent material of the evidentiary anthrax spore powder, *i.e.*, the evidentiary material came from a derivative growth of RMR-1029" (USDOJ, 2010, p. 28).	The results of the genetic analyses of the repository samples were consistent with the finding that the spores in the attack letters were derived from RMR-1029, but the analyses did not definitively demonstrate such a relationship. The scientific data alone do not support the strength of the government's repeated assertions that that "RMR-1029 was conclusively identified as the parent material to the anthrax powder used in the mailings" (USDOJ, 2010, p. 20), nor the role suggested for the scientific data in arriving at their conclusions, "the scientific analysis coordinated by the FBI Laboratory determined that RMR-1029, a spore-batch created and maintained at USAMRIID by Dr. Ivins, was the parent material for the anthrax used in the mailings" (USDOJ, 2010, p. 8).[b]	Finding 6.2

TABLE S-1 Continued

FBI conclusions	DOJ conclusions	Committee comment	Relevant report finding/section
	"If Dr. Ivins prepared his submission to the repository in accordance with the protocol, that submission could not miss all four of the morphological variants present in RMR-1029" (USDOJ, 2010, p. 79).	The genetic evidence that a disputed sample submitted by the suspect came from a source other than RMR-1029 was weaker than stated in the Department of Justice *Amerithrax Investigative Summary*.	Finding 6.4
	"…the only complete genetic match to the evidence comes from RMR-1029 and its offspring" (USDOJ, 2010, p. 29).	The scientific data generated by and on behalf of the FBI provided leads as to a possible source of the anthrax spores found in the attack letters, but these data alone did not rule out other sources.	Finding 6.5

continued

TABLE S-1 Continued

FBI conclusions	DOJ conclusions	Committee comment	Relevant report finding/ section
Out of the 1,059 viable samples from various stocks collected in the FBIR during the investigation, 8 contained all 4 of the polymorphisms; 2 contained 3 of the 4 polymorphisms; and a few contained 1 or 2 of the mutations. (FBI, 2009).		The FBI and contract scientists appropriately recognized that the mutations in the letter isolates provided information that might help identify the source of the *B. anthracis* used in the attacks, developed appropriate assays for four of these mutations, and created and screened a repository of Ames strain samples. Based on the results of that screening, FBI scientists appropriately concluded that the majority of repository samples contained none of the four mutations, although 50 of the samples contained one of the four mutations and 10 samples had three or all four mutations (the numbers with one or more mutations are higher if one includes samples that were excluded in the FBI's *Statistical Analysis Report*). However, features of the repository, including unknown provenance, possibly multiple samples from the same flask, the history of sharing and mixing of stocks presented investigative challenges.	Finding 6.1 discussion

TABLE S-1 Continued

[a]Note that this was the final conclusion of the scientific investigators. An initial finding by the Armed Forces Institute of Pathology (AFIP) found, upon gross examination, that the spores exhibited a silicon signal and sometimes exhibited an oxygen signal. Subsequent studies conducted by Sandia National Laboratories (as described in Chapter 4 of this report) determined that the silicon was localized to the spore coat within the exosporium—that is, it was incorporated into the cell as a natural part of the cell formation process. The USAMRIID scientist who first reviewed the AFIP results and made statements regarding the presence of silicon and possible weaponization retracted those earlier statements.

[b]See for example, "As noted above, based on advanced genetic testing combined with rigorous investigation, the FBI concluded that RMR-1029 is the parent material of the evidentiary anthrax spore powder, *i.e.*, the evidentiary material came from a derivative growth of RMR-1029." (p. 28) "As described in detail above, over time, genetic analysis determined that one of Dr. Ivins's Ames cultures, RMR-1029—the purest and most concentrated batch of Ames spores known to exist—was the parent to the evidentiary material used in the anthrax mailings." (p.79)

1

Introduction

1.1 BACKGROUND

In fall 2001, shortly after the September 11 attacks in New York City and Washington, D.C., U.S. citizens experienced a second set of attacks, this time involving the bacterium *Bacillus anthracis* (*B. anthracis*, or more simply, anthrax) placed in at least four and possibly five letters and sent through the mail. From October 4 to November 20, 2001, 22 cases of anthrax were identified—11 inhalational and 11 cutaneous. Five of the inhalational cases were fatal (Jernigan et al., 2002). Twenty infected individuals contracted anthrax as mail handlers or at worksites where contaminated mail was processed or received. Two victims who died from the infection had no known contact with any of the worksites in question. An additional 31 people tested positive for exposure to *B. anthracis* spores; approximately 32,000 individuals initiated antibiotic prophylaxis (Centers for Disease Control and Prevention [CDC], 2001a; Jernigan et al., 2002).

Over the course of its investigation, known by the case name "Amerithrax," the Federal Bureau of Investigation (FBI) devoted 600,000 investigator work hours to the case and assigned 17 Special Agents to a Task Force, along with 10 U.S. Postal Inspectors. The investigation spanned six continents; involved over 10,000 witness interviews, 80 searches, 26,000 email reviews, and analyses of 4 million megabytes of computer memory; and resulted in the issuance of 5,750 grand jury subpoenas (U.S. Department of Justice [DOJ], 2010, p. 4). Additionally, 29 government, university, and commercial laboratories assisted in conducting the scientific analyses that were a central aspect of the investigation (U.S. Department of Justice [USDOJ], 2010, p. 4).

The investigation also accelerated the development of a nascent scientific field, called microbial forensics, involving a series of laboratory tests to pinpoint the genetic identity of a microbial agent used for nefarious purposes. This field grew out of the multidisciplinary areas of genomics, microbiology, and forensics, among others. The development and application of microbial forensics became an essential part of the scientific investigation in the hands of

FBI investigators, who combined it with physicochemical analyses to narrow their search for the source of the anthrax used in the attacks.[1]

In 2008, seven years into the Amerithrax investigation, the FBI asked the National Research Council of the National Academy of Sciences (NAS) to conduct an independent review of the scientific approaches used during the investigation of the 2001 *B. anthracis* mailings (see Box 1-1).

During the course of the NRC committee's deliberations, the DOJ announced on February 19, 2010, that it was closing the case based on its conclusion that Dr. Bruce Ivins, a scientist at the U.S. Army Medical Research Institute of Infectious Diseases (USAMRIID), had alone perpetrated the anthrax attacks. Dr. Ivins died on July 29, 2008 after taking an overdose of over-the-counter medications.

The committee carried out its work mindful of the need to identify lessons that could be learned for future investigations in which science might play an important role.

1.2 CHRONOLOGY OF EVENTS OF FALL 2001

Public health officials in Florida announced on October 4, 2001, that Robert Stevens, a photo editor at American Media, Inc. (AMI) in Boca Raton, had inhalational anthrax. This was the first reported case of inhalational anthrax in the United States in almost 25 years. After one of Stevens's coworkers, Ernesto Blanco, also fell ill and was diagnosed as having contracted anthrax, environmental assessments were made of the AMI facility. These assessments revealed *B. anthracis* contamination and postexposure prophylactic treatment was administered to AMI employees. No contaminated letter was ever found; it is thought to have been discarded after being opened (CDC, 2001a). A timeline of this and subsequent events is presented in Table 1-1.

Less than two weeks later, additional cases of apparent anthrax exposure began to appear in New York City. These cases indicated the possible source of the exposure as most of those infected had come into contact with letters containing a powder. The New York letters addressed to Tom Brokaw of NBC News and the *New York Post* had a Trenton, New Jersey, postmark dated September 18, 2001. Sampling of U.S. Postal Service drop boxes in the Trenton area found anthrax spores in only one mailbox, on Nassau Street in Princeton (see Chapter 3).

A second wave of mailings caused additional cases of anthrax. Two more anthrax letters bearing the same Trenton postmark, but dated October 9, 2001, were addressed to Democratic U.S. Senators Tom Daschle of South Dakota and

---

[1] In 2008 the National Bioforensic Analysis Center was established in the Department of Homeland Security's National Biodefense Analysis and Countermeasure Center to assist in microbial forensics investigations.

BOX 1-1
Charge to the Committee

The NRC was asked by the FBI to conduct an independent review of the scientific approaches used during the investigation. The official charge to the committee stated:

> An ad hoc committee with relevant expertise will evaluate the scientific foundation for the specific techniques used by the FBI to determine whether these techniques met appropriate standards for scientific reliability and for use in forensic validation, and whether the FBI reached appropriate scientific conclusions from its use of these techniques. In instances where novel scientific methods were developed for purposes of the FBI investigation itself, the committee will pay particular attention to whether these methods were appropriately validated. The committee will review and assess scientific evidence (studies, results, analyses, reports) considered in connection with the 2001 *Bacillus anthracis* mailings. In assessing this body of information, the committee will limit its inquiry to the scientific approaches, methodologies, and analytical techniques used during the investigation of the 2001 *B. anthracis* mailings.
>
> The areas of scientific evidence to be studied by the committee include, but may not be limited to:
>
> 1. genetic studies that led to the identification of potential sources of *B. anthracis* recovered from the letters;
> 2. analyses of four genetic mutations that were found in evidence and that are unique to a subset of Ames strain cultures collected during the investigation;
> 3. chemical and dating studies that examined how, where, and when the spores may have been grown and what, if any, additional treatments they were subjected to;
> 4. studies of the recovery of spores and bacterial DNA from samples collected and tested during the investigation; and
> 5. the role that cross contamination might have played in the evidence picture.
>
> The committee will necessarily consider the facts and data surrounding the investigation of the 2001 *Bacillus anthracis* mailings, the reliability of the principles and methods used by the FBI, and whether the principles and methods were applied appropriately to the facts. The committee will not, however, undertake an assessment of the probative value of the scientific evidence in any specific component of the investigation, prosecution, or civil litigation and will offer no view on the guilt or innocence of any person(s) in connection with the 2001 *B. anthracis* mailings, or any other *B. anthracis* incidents.

TABLE 1-1 Timeline of Key Events in the Anthrax Mailings Case

2001	**September:** Letters containing anthrax spores are mailed to news organizations in New York (ABC, CBS, NBC, and the *New York Post*) and Florida (American Media, Inc.). While only two letters are actually recovered (one addressed to the *New York Post* and the other to Tom Brokaw at NBC), the existence of other letters is inferred from the pattern of infection (Piggee, 2008; Ember, 2006).

September 18: Postmark date on the *Post* and Brokaw anthrax letters. The postmark indicates that the letters were mailed from Trenton, New Jersey (Cole, 2009, p. 89).

October: Letters containing anthrax spores are mailed to U.S. Senators Thomas A. Daschle and Patrick Leahy in Washington, D.C. The FBI begins an investigation—code-named Amerithrax—into the mailings (Piggee, 2008).

October 4: Robert Stevens, a photo editor working for American Media, Inc., in Boca Raton, Florida, is diagnosed with inhalational anthrax, believed to have been contracted as a result of contamination of his workplace by an anthrax mailing. The diagnosis, initially made by a physician-microbiologist at the hospital where Stevens received care, was then confirmed at the Florida State Laboratory in Jacksonville and the U.S. Centers for Disease Control and Prevention (Traeger, 2002).

October 5: Robert Stevens dies from inhalational anthrax. He is the first of five persons to die of the illness. In total, 11 individuals are believed to have contracted inhalational anthrax as a result of the mailings (Ember, 2006; Cole, 2009, p. 197).

October 9: Postmark date on the Daschle and Leahy anthrax letters. The postmark also indicates that the letters were also mailed from Trenton, New Jersey (USDOJ, 2010).

October 12: The FBI recovers the Brokaw letter (USDOJ, 2010, p. 4). A case of cutaneous anthrax is confirmed in Erin O'Connor, an assistant to Tom Brokaw. She is the first of 11 persons believed to have contracted cutaneous anthrax as a result of the anthrax mailings (Cole, 2009, p. 54).

October 15: The Daschle letter is opened in the Senator's office in the Hart Senate Office Building (Cole, 2009, p. 89).

October 16, 17: The Hart Senate Office Building and other U.S. Senate and House office buildings are closed (Ember, 2006).

October 18: The U.S. Postal Service's Trenton Processing and Distribution Center in Hamilton Township, New Jersey, is closed for anthrax testing (Cole, 2009, p. 92). On the same day, in a joint announcement with Postmaster General Jack Potter, FBI Director Robert Mueller offers a $1 million reward for "information leading to the arrest and conviction for terrorist acts of mailing anthrax" (FoxNews, 2001).

October 19: The *New York Post* letter is discovered and recovered (USDOJ, 2010, p. 4).

October 21: Thomas L. Morris, Jr., a postal worker at the Washington, D.C., Brentwood Mail Processing and Distribution Center, which serviced Capitol Hill, is the second person to die from inhalational anthrax believed to have been contracted as a result of the anthrax mailings (Cole, 2009, p. 65). The Brentwood Mail Processing and Distribution Center is closed the same day (Cole, 2009, p. 75).

October 22: Joseph P. Curseen, Jr., a postal worker at the Washington, D.C., Brentwood Mail Processing and Distribution Center, is the third person to die from inhalational anthrax believed to have been contracted as a result of the anthrax mailings (Cole, 2009, p. 65).

TABLE 1-1 Continued

	October 31: Kathy T. Nguyen, a hospital worker at the Manhattan Eye, Ear, and Throat Hospital, is the fourth person to die from inhalational anthrax believed to have been contracted as a result of the anthrax mailings (Cole, 2009, p. 5).
	November 16: In a joint operation, the FBI, U.S. Environmental Protection Agency, and U.S. Postal Inspection Service discover the Leahy letter in a bag of unopened mail (Piggee, 2008).
	November 21: Ottilie Lundgren, an elderly woman in Oxford, Connecticut dies from inhalational anthrax. She is the last person to die from inhalational anthrax believed to have been contracted as a result of the mailings (Cole, 2009, p. 108).
	December: The Leahy letter is opened and examined at the U.S. Army Medical Research Institute for Infectious Diseases (USAMRIID) at Fort Detrick in Frederick, Maryland (Cole, 2009, p. 90).
	December 31: The Dirksen Senate Office Building, which is connected to the Hart Senate Office Building by underground corridors, is reopened (*New York Times*, 2002).
2002	**January 23:** The Hart Senate Office Building is reopened. On the same day, the FBI increases the reward for help in solving the case to $2.5 million (Gallucci-White, 2008, p. 8).
	June: Officials say the FBI is "scrutinizing 20 to 30 scientists who might have had the knowledge and opportunity to send the anthrax letters" (Gallucci-White, 2008, p. 8).
	August 6: Attorney General John Ashcroft publicly names Steven J. Hatfill, a former USAMRIID scientist and biodefense expert, as a "person of interest" in the Amerithrax investigation. Hatfill would be cleared in 2008 (Freed, 2010).
2003	**March:** Anthrax decontamination begins at the American Media, Inc., building in Boca Raton, Florida, where Robert Stevens worked (BioOne, 2005).
	June: Searching for evidence related to the anthrax mailings, the FBI drains a pond in Frederick, Maryland. Nothing suspicious is found (Cole, 2009, p. 195).
	August: Steven Hatfill sues Attorney General John Ashcroft and other government officials, accusing them of using him as a scapegoat and demanding that his name be cleared (*Washington Post*, 2008, p. 11).
	December 22: The U.S. Postal Service's Brentwood Mail Processing and Distribution Center is reopened (USDOJ, 2010, p. 3).
2005	**March 14:** The U.S. Postal Service's Trenton Processing and Distribution Center in Hamilton Township, New Jersey, is reopened (USDOJ, 2010, p. 3).
2007	**February 8:** Federal environmental experts determine that the former American Media, Inc., building in Boca Raton, Florida, has been cleared of anthrax spores (Sarmiento, 2007).
2008	**June:** The federal government awards Steven Hatfill $5.82 million to settle his violation of privacy lawsuit against the Department of Justice (DOJ) (Freed, 2010; *Washington Post*, 2008).

continued

TABLE 1-1 Continued

> **July 29:** USAMRIID microbiologist Bruce E. Ivins commits suicide as the FBI is about to file criminal charges against him for his role in the anthrax mailings (CBS News, 2008).
>
> **August 8:** DOJ officially clears Steven Hatfill of involvement in the anthrax mailings (*Washington Post*, 2008).
>
> **August 18:** The FBI holds two press briefings, one for scientific media and one for general media, to describe "the body of powerful evidence" that allowed the FBI to conclude that it had "identified the origin and perpetrator of the 2001 *Bacillus anthracis* mailing" (FBI, 2008).
>
> **September 17:** FBI Director Robert Mueller testifies before the Senate Judiciary Committee at a hearing entitled "Oversight of the Federal Bureau of Investigation." At the hearing, Mueller states that the FBI was seeking an independent review of the scientific evidence in the anthrax mailings case. "Because of the importance of the science to this particular case and perhaps cases in the future," he says, "we have initiated discussions with the National Academy of Sciences" to "undertake a review of the scientific approach used during the investigation" (Temple-Raston, 2008).

2010 **February 19:** DOJ, the FBI, and the U.S. Postal Inspection Service formally conclude the investigation into the 2001 anthrax attacks and the Department of Justice issues an *Amerithrax Investigative Summary*. In the summary, DOJ concludes that "Evidence developed from [the] investigation established that Dr. [Bruce] Ivins, alone, mailed the anthrax letters" (USDOJ, 2010, p. 1).

Patrick Leahy of Vermont. The letter addressed to Senator Daschle was opened by a member of the Senator's staff on October 15. After discovering the Daschle letter, the House and Senate Office Buildings were closed for environmental assessment and decontamination. The U.S. Postal Service suspended mail service to the U.S. Capitol and closed the Hamilton, New Jersey, postal center where the four recovered letters had been processed.

Postal officials subsequently determined that two contaminated envelopes were processed at the U.S. Postal Service Processing and Distribution Center in Washington, D.C. (the Brentwood facility) on October 12. Exposure to spores from the unopened envelopes at the postal facilities went undetected until after the implicated envelope was opened at the Hart Senate Office Building. On October 21, officials closed the Brentwood facility after a postal worker was diagnosed with an anthrax infection. Several workers at the postal facility that processed the letter fell ill with inhalational anthrax, and two eventually died.

On November 16, 2001 FBI officials, U.S. Postal investigators, and U.S. Environmental Protection Agency hazardous material personnel found an unopened letter addressed to Senator Leahy that appeared to contain anthrax. The letter, with an October 9, 2001, Trenton, New Jersey postmark, was located in one of more than 230 drums—containing 642 bags of unopened mail sent to Capitol Hill—that had been sequestered since the discovery of the anthrax letter mailed to Senator Daschle (Beecher, 2006). The letter had a Greendale School return address, block handwriting, and other characteristics similar to the Daschle letter. A misread zip code caused the Leahy letter to be misdirected to the State Department mail annex in Sterling, Virginia, where a postal worker contracted inhalational anthrax.

The anthrax in the Senate letters was a highly refined dry powder consisting of about one gram of nearly pure spores, as determined in subsequent laboratory analyses (see Chapter 4). The preparation was thus more potent than the material in the first (New York) set of mailings.

By the beginning of December 2001, it appeared that the mailings had ended, as no additional letters had been discovered and no further cases were identified. But it was clear that more was required than a public health response by CDC. The attacks warranted a major law enforcement investigation led by the FBI, in which science would play a leading role. Identifying the source of the letter materials could lead to the person or persons responsible for the attacks. Key questions focused on the contents of the letters, how, where, and when the materials in the letters might have been produced, whether the material in all the evidence collected was identical, whether the material had been produced in such a manner as to be more easily dispersible, whether it had any distinguishing physical or chemical properties of value in determining the source, and whether its biological characteristics could provide leads to its origins.

1.3 BRIEF SUMMARY OF THE FBI'S SCIENTIFIC INVESTIGATION

During its investigation of the anthrax mailings, the FBI worked with other federal agencies to coordinate and conduct scientific analyses of the spore powders recovered from the letters, environmental samples, clinical samples, and samples collected from laboratories that might have been the source of the letter-associated spores. The agency relied on external experts, including some who had previously developed tests to differentiate among strains of *B. anthracis*.

Early in the investigation the spores in the letters, as well as environmental and clinical isolates, were identified as the "Ames strain" of anthrax. This strain was originally isolated from a dead cow in Texas in 1981 and shipped to USAMRIID in Frederick, Maryland. Over time it was shared with research and development laboratories around the world. Thus, the identification of the strain of *B. anthracis* used in the mailings was insufficient to identify its source, although it narrowed the possibilities considerably. The evidence had to be examined for additional unique and distinguishing features that could then be compared to samples obtained from laboratories holding the Ames strain as a means to narrow the search for the possible source material, and perpetrator(s).

The FBI subpoenaed samples from laboratories known to have Ames strain *B. anthracis* and collected them in an FBI Repository (FBIR) that ultimately included 1,070 samples from 20 laboratories—17 domestic and 3 international (in Canada, Sweden, and the United Kingdom). In addition, the FBI and partners within the Intelligence Community collected environmental samples from an undisclosed overseas site at which they had reason to suspect activities by a terrorist group in producing anthrax. Although cultures of these samples did not produce *B. anthracis*, molecular analysis provided inconsistent evidence for the presence of *B. anthracis* Ames strain DNA in some samples (see section 3.4.3)

Scientists from the Department of Defense examined the spore materials in the letters and identified several variants in the samples based on their colony morphology.[2] With support from the National Institutes of Health, the National Science Foundation, and other government agencies, FBI scientists worked with the Institute for Genomic Research (TIGR) to identify several genetic mutations associated with the altered appearance of the cultured variants found in the letters (see Chapter 5 for an extensive discussion of this work).

FBI investigators contracted the assistance of four laboratories to develop highly specific molecular-genetic assays to detect four specific mutations found in the evidence. These mutation detection assays were used in the examination of the samples in the FBIR, as described in Chapters 5 and 6.

The analysis of samples in the FBIR led the FBI to focus attention on a particular spore-containing flask at USAMRIID known as RMR-1029. The analysis of the repository samples and the bacteria in this flask is described in Chapter 6.

In addition, analytical approaches such as scanning and transmission electron microscopy, energy-dispersive X-ray analysis, carbon dating by accelerator mass spectrometry, and inductively coupled plasma-optical emission and mass spectrometry were used to determine the chemical and elemental profiles of the spore powders (see Chapter 4). These studies were done to determine when the anthrax preparation might have been made, whether there were contaminants or trace elements that would provide a clue to the production location or materials used, and whether there was evidence of an effort to deliberately include additives to improve dispersal of the anthrax.

1.4 SUMMARY OF FBI AND DOJ SCIENTIFIC CONCLUSIONS

The scientific analyses led the FBI and DOJ to draw a number of conclusions (see Table S-1 in the Summary). The committee found it challenging, however, to identify the FBI's definitive conclusions because those provided publicly by DOJ in its briefings and investigative summary and those provided by FBI officials in oral presentations to the committee varied. For the purposes of this

---

[2] Morphological variants are observable physical or biochemical characteristics of an organism. These characteristics are determined by both genetic makeup and environmental influences.

report, the committee's analyses are based on the scientific conclusions provided by the FBI to the committee on September 24, 2009 (left-most column of Table S-1), those issued publicly by DOJ on February 19, 2010, when it closed the case (FBI, 2010) (column second from the left in Table S-1), and those provided by Louis Grever, Edward Montooth, and Rachel Lieber on January 14, 2011 (FBI/USDOJ, 2011).

1.5 COMMITTEE PROCESS

Under the terms of the NRC contract with the FBI, the FBI initially provided two boxes containing approximately 9,000 pages of materials to the committee, and then in December 2010, the FBI gave the committee an additional 641 pages related to the scientific investigations undertaken by the FBI and by various external experts working at the behest of the Bureau during the course of the anthrax investigation. Throughout the NRC study process these materials were covered by FOIA Exemption 7, "law enforcement sensitive," and were not publicly available. Upon release of this report, as specified in the contract, these documents have been deposited in the NRC Public Access File.[3]

Documents were initially delivered in two batches containing reports of the scientific analyses (see the Index of Documents Provided by the Federal Bureau of Investigation for a listing).[4] The first batch included technical review panel reports, laboratory analytical test reports and results pertaining to Ames strain identification, carbon dating, stable isotope analysis, agar and heme analysis, and assay development, and published papers. Batch two included materials regarding genetic diversity and phylogenetic characterization of *B. subtilis* (another bacterial species); repository screening and molecular analysis of pathogen strains and isolates and genetics of the A1, A3, B, D, and E mutations found in the evidence; statistical analysis; cross contamination; and chemical and physical properties of the spore powders. The third batch of documents received in December 2010 contained reports of scientific review meetings and some additional information about sample collection, laboratory notebooks, and reports of investigations of individuals. Additional documents were provided by the FBI at the committee's request throughout the study; these documents are listed in the Index of Documents Provided by the Federal Bureau of Investigation under the heading Supplemental Documents.

No written explanatory materials were provided with these documents that would fully inform the committee as to why the analyses were done and how these

---

[3] The public can gain access to these materials by contacting the NRC Public Access Records Office.

[4] In this report, the principal documents received from the FBI are referenced according to the following convention: "FBI Documents, B*M*D*" where B = Batch, M = Module, D = Document, and * = Number. Thus "FBI Documents, B1M1D1" would refer to the first document in the first module of the first batch of materials received from the FBI.

documents contributed to the FBI investigations and conclusions. The material regarding analyses of the FBIR specimens was coded, often with different numbers for the same sample set. Consequently, the committee spent a considerable amount of time sorting through and attempting to interpret the available materials before it could begin to evaluate the science and consider the scientific conclusions. In addition, much of the information provided to the committee was compartmentalized and sections of some documents were redacted.

When the committee posed questions to the FBI for clarification, the agency was always responsive; however, responses to questions were sometimes minimal or terse, or were deflected as intruding into the criminal investigation and beyond the purview of the committee despite the committee's explanation that the questions were of a scientific nature. Some of these responses may reflect tension between the scope of the scientific review expected by the FBI and the committee's interpretation of its charge. In summary, the FBI provided some of the primary information related to the scientific analyses and was generally responsive to committee questions, but early on it was difficult for the committee to ascertain details of what was done in the course of some of the FBI scientific work, the identity of some of the samples analyzed, and the relationships among the samples in the repository.

In addition to materials provided directly by the FBI to the committee, FBI officials also briefed the committee on several occasions. Some of these briefings were done in open session, while others were conducted in closed sessions covered by FOIA Exemption 7. In these closed sessions the committee heard from a number of DOJ/FBI personnel including: John Fraga, Christian Hassell, Louis Grever, Edward Montooth, and Rachel Leiber. FBI consultant Ranajit Chakraborty (University of Cincinnati and currently University of North Texas Health Sciences Center) and Daniel Martin (Dugway Proving Ground) also briefed the committee in closed session. In addition, the committee heard from a number of other experts: Bruce Budowle (formerly FBI; University of North Texas Health Sciences Center); Rita Colwell (University of Maryland and Johns Hopkins University); Claire Fraser-Liggett (The Institute for Genomic Research and University of Maryland School of Medicine); Hank Heine (formerly USAMRIID); Congressman Rush Holt; Paul Keim (Northern Arizona University); Joseph Michael (Sandia National Laboratories); Steven Schutzer (University of Medicine and Dentistry of New Jersey); Jennifer Smith (formerly FBI; BIOFOR Consulting); Patricia Worsham (USAMRIID); and Peter Weber (Lawrence Livermore National Laboratory).

In conducting its review, the committee focused on the biological, physical, and chemical sciences applied to evidentiary materials. The committee was not charged to consider or evaluate any of the traditional forensic science methods and techniques used (e.g., hair, fiber, fingerprint, or handwriting analysis) (NRC, 2009a) nor did it consider any of the psychological or behavioral sciences, such as linguistics, used by the FBI in its investigation.

INTRODUCTION   35

The committee met seven times in person in open and closed sessions and continued its deliberations by conference call. Closed sessions were reserved for review of law enforcement materials (FOIA Exemption 7) relevant to the FBI investigation, committee analyses, deliberations, and report drafting. Public sessions were convened to gather information from the scientific community about various aspects of the scientific investigation or areas of scientific research relevant to the matters at hand. DOJ closed its investigation of the anthrax mailings after the committee's fourth meeting.

In November 2010, the FBI contacted the National Academies and requested the opportunity to provide the committee with additional materials and another briefing. The committee subsequently received and reviewed the third batch of materials, an additional 641 pages of documents, and met for one final briefing with FBI and DOJ officials in mid-January 2011.

1.6 ISSUES FOR CONSIDERATION IN READING THIS REPORT

The FBI's anthrax investigation involved the development and use of modern science in an attempt to solve a crime committed with a biological agent. The use of science in legal investigations is not new. Science is called on to answer questions, for example, about the safety of drugs, risks from exposure to environmental toxins, and identification of DNA from a rape or murder victim. Yet science and the judicial system do not always have an easy relationship because of differences in culture and overall objectives. The scientific process takes time, raising questions and seeking answers, and challenging and revising accepted theories and notions until new hypotheses are generated. The judicial system, on the other hand, aims to settle disputes with the information available at a point in time. It typically does not have the opportunity to conduct another study and wait for complete information. Scientific investigation usually is a more open-ended endeavor than a legal or criminal investigation as scientists acknowledge appropriate degrees of uncertainty—both small and large—in their investigations and are inspired to do future work on the questions of interest, yielding more certainty and more information. In contrast, the justice system, to be effective, requires decisions to be made rather than deferred, and thus scientific uncertainty has to be weighed in light of all other evidence. Tolerance for scientific uncertainty may or may not be tempered by the strength of other, nonscientific evidence.

As demonstrated in this investigation, the FBI used science in two different ways: 1) to identify and analyze evidence using methods that are acceptable for presentation in the courtroom; and 2) to identify leads for a criminal investigation. In the latter case, the science *per se* is not intended to be presented in the courtroom but it may provide leads to inform and direct the law enforcement investigation. In either case, the science must be conducted correctly and performed at a high level of scientific standards.

The committee recognized that forensic science is the application of scientific methods to matters of interest to the judicial system and must, therefore, consider the norms of both science and the law (NRC, 2009a, Chapter 3). The committee also recognized that sometimes pressing national interest or security concerns, such as those present during this investigation, demand that newly emerging methods be applied to the assessment of forensic evidence even before those methods have been widely adopted or validated by peer review in the forensic and scientific communities. It should be noted that future biological attacks will probably pose greater challenges than did this attack: the agent may be a member of a species with a more complex and poorly understood population structure, the agent may be genetically modified in a manner that further obscures its origin, or a sample of the attack material may not be readily available (as it was in this case). This last possibility may mean that environmental or clinical samples, with their inherent added challenges, will have greater importance in a future investigation.

National security concerns and the pressures of an ongoing criminal investigation may require that the collection of samples and their evaluation be carried out under circumstances of secrecy that limit the capacity of outside observers to assess the validity of the forensic interpretations. Such circumstances pose special challenges in which the optimal application and evaluation of scientific methods may in some instances run counter to security interests. The committee faced this tension between science and security in its deliberations.

In the end, the committee considered the facts and data of the scientific investigation, the reliability of the principles and methods used by the FBI, whether the principles and methods were applied appropriately to the facts, and the conclusions related to these efforts. The committee does not, however, offer a view on the guilt or innocence of any person(s) in connection with the 2001 *B. anthracis* mailings or any other *B. anthracis* incidents.

1.7 ORGANIZATION OF THE REPORT

Based on its review of the materials provided, the committee developed the findings presented in this report. The report is organized to provide background on the scientific characteristics of *B. anthracis* (Chapter 2); describe and review the procedures used in the early stages of the investigation concerning the collection of evidence and its processing and preservation, as well as the creation of a repository of *B. anthracis* samples collected from around the world for comparative and investigative purposes (Chapter 3); review and assess the physicochemical analyses of the anthrax evidence (Chapter 4); review and assess the biological characteristics of the material in the letters (Chapter 5); and review and assess the analyses and results of the FBI's comparison of the evidentiary material against the samples in the FBI Repository (Chapter 6). The committee's findings, analysis, and recommendations can be found in Chapters 3 through 6.

2

Biology and History of *Bacillus anthracis*

2.1 INTRODUCTION

Bacillus anthracis (*B. anthracis*) is a Gram-positive rod-shaped bacterium that is the causative agent of the disease anthrax. *B. anthracis* rods typically have dimensions of approximately 1 μm by 4 μm and may occur in chains resembling "boxcars" when observed under a microscope. The word "anthrax" is derived from the Greek word for "coal" or "black" and was applied to this disease because patients with the cutaneous form of the disease often exhibit black skin lesions. The disease has been known since antiquity. In the 19th century, the demonstration by Robert Koch of *B. anthracis* as the cause of anthrax was a cornerstone of proving the germ theory of disease.

This chapter summarizes historical, biological, and phylogenetic features of *B. anthracis,* and in particular the Ames strain, as well as clinical features of anthrax, many of which proved highly relevant as the events surrounding the mailing of the *B. anthracis* spore-laden letters unfolded in fall 2001.

2.2 THE BIOLOGY OF *B. ANTHRACIS*

B. anthracis spores are the infectious agent for all forms of anthrax. In contrast, *B. anthracis* vegetative cells are noninfectious in animal models (Mock and Fouet, 2001). The structural characteristics and environmental resistance of the *B. anthracis* spore are key to its avoidance of the host innate immune response during the initial stages of infection (Cybulski et al., 2009; Mock and Fouet, 2001). Relative to the majority of bacterial vegetative cells, spores like those produced by *B. anthracis* are highly resistant to a variety of commonly lethal treatments, including dehydration, elevated temperature, UV irradiation, numerous toxic chemicals, and enzymatic digestion by proteases and lysozymes (Setlow, 2006). In the dormant state, spores can retain viability for decades and likely for centuries (Nicholson et al., 2000); however, upon arrival in a hospitable environment, the spores can germinate and resume rapid growth within hours (Moir, 2006; Setlow, 2003). The stability of spores, with accompanying ease of storage and transport, is also a major factor in their potential utility as a biological weapon.

The resilience of *B. anthracis* spores derives from their unique physical and structural characteristics. The spore cytoplasm or "core" is relatively dehydrated and contains high concentrations of certain low-molecular-weight solutes (Setlow, 2006). These conditions result in complete metabolic dormancy and extreme protein stability. The low core water content of the spore correlates with a high spore density, a property that is commonly exploited for spore purification. Centrifugation through density gradients of diatrizoate and meglumine is a widely used purification method (Tamir and Gilvarg, 1966; Nicholson and Setlow, 1990).

Surrounding the spore core are a membrane and a specialized peptidoglycan cell wall, the "cortex." These structures play key roles in limiting movement of water and solutes into and out of the core, maintaining the dehydrated dormant state (Setlow, 2006). The membrane and cortex have not been cited as significant sites of mineral association nor found to play roles in direct spore interactions with external surfaces.

The outermost layers of the spore, the coat and exosporium, are the primary sites of interaction with the host and with other surfaces and are important factors in the resistance of spores to certain enzymatic and chemical treatments (Driks, 2009; Henriques and Moran, 2007; Setlow, 2006). These structures are composed predominantly of protein and glycoprotein. In some spore-forming bacteria, the spore coats have been shown to be sites of association of minerals (Johnstone et al., 1980; Mann et al., 1988; Stewart et al., 1980, 1981; Hirota et al., 2010). These outermost integuments can play major roles in determining spore adherence to surfaces (Bozue et al., 2007; Brahmbhatt et al., 2007) and may affect electrostatic properties and aggregation with other spores or particles, all of which will affect spore dispersal and infectivity. Efforts at spore aerosolization and dispersal have been pursued through modification of the spores' surface structures and properties (Swartz, 2001).

The fact that *B. anthracis* can exist in the environment as a dormant, highly stable spore may have evolutionary significance. While a dormant environmental state may help explain observations that populations of *B. anthracis* have much less genetic variation than many other bacterial species (Van Ert et al., 2007a), other factors may also contribute, including the likely recent origin of this species and the possibility of limited means for horizontal gene transfer.

2.3 CLINICAL ASPECTS OF ANTHRAX

Anthrax is generally a disease of herbivores (e.g., cattle, sheep, horses), which acquire the infection by grazing on contaminated soils. Anthrax spores are highly resistant to environmental insults. These spores allow the bacterium to survive for long periods of time in soil, its natural reservoir. The disease occurs worldwide, and there are occasional outbreaks of anthrax in livestock in the United States and Canada.

The course and outcome of human anthrax depend on dose and on whether the infection is acquired via the skin, gastrointestinal tract, or inhalation. Most cases of human anthrax involve skin lesions (cutaneous anthrax), whereby the infection is usually acquired as a result of handling infected animal hides or wool, leading to contamination of skin abrasions with *B. anthracis*. Cutaneous anthrax is the least lethal form of the disease, but still can cause significant mortality of up to 20 percent (Atlas, 2002) if not treated with antimicrobial therapy. Ingestion of food, such as meat contaminated with *B. anthracis*, can produce gastrointestinal anthrax, which is a serious disease with 25 to 60 percent fatality. This form of the disease is extremely rare in developed countries. The most fulminant and lethal manifestation of anthrax is due to inhalation of *B. anthracis* spores, causing a highly fatal disease. Inhalational anthrax is generally rare and is observed mainly in individuals who work with animal skins. In the 19th century, it was known as "woolsorter's disease" and is believed to have been the first documented occupational illness (Leffel and Pitt, 2006). Inhalational anthrax in individuals not likely to have suffered occupational exposure, however, can be a sign of a biological attack with *B. anthracis* spores.

Inhalational anthrax is a rapidly progressive disease with high mortality and morbidity even when treated with antimicrobial therapy. Anthrax spores germinate when placed in blood or other human or animal tissues that provide a nutrient-rich environment (Inglesby et al., 2002). Upon inhalation, the host's macrophages (a type of immune system cell) attack and ingest the spores, which are protected from these host cells by the spore coat. The macrophages unwittingly transport the spores to lymph nodes in the respiratory system (Liddington, 2002), where they are released. The lymph nodes provide sufficient nutrients to allow the spores to germinate and begin to proliferate. Proliferation in the lymphatic system in turn allows the bacteria in their "vegetative" state to spread into the blood stream and be disseminated to multiple organs.

The genetic determinants of virulence in *B. anthracis* reside primarily on two large plasmids, which are extrachromosomal DNA molecules. These plasmids are known as pXO1 and pXO2, and they contain genes that encode for anthrax toxin and a poly-D-glutamate capsule, respectively. Anthrax toxins are composed of a combination of three proteins that work together: protective antigen (PA), edema factor (EF), and lethal factor (LF). When PA combines with EF and LF, toxicity to host cells and a buildup of fluids (edema) in infected tissues (e.g., the lungs) are produced. EF is an adenylate cyclase enzyme that promotes the accumulation of cyclic adenosine monophosphate (AMP), producing a loss of cellular regulation of water and ion metabolism. LF is a metallopeptidase that cleaves proteins of signal transduction pathways, resulting in profound effects that range from cell death to interference with the cellular functions necessary for mounting appropriate immune responses. The effects of the anthrax toxin can kill an infected patient even after the bacteria in the patient's body have been killed by antibiotics (Bouzianos, 2009).

In addition to the PA, EF, and LF toxin components, *B. anthracis* produces other cytotoxic enzymes, such as anthrolysin O, that contribute to pathogenesis. The combination of the protein toxins, which undermine the host's defenses and interfere with cellular function, and the polymerized amino acid capsule, which protects the bacterium from phagocytic cells, represents a particularly lethal mix. Antibodies that neutralize the toxins or promote phagocytosis of encapsulated cells can confer significant protection to the host. The vaccine currently in use for the prevention of anthrax functions by eliciting neutralizing antibody responses to anthrax toxin. *B. anthracis* mutants that lack either toxin production or capsules are attenuated for virulence and can be used in vaccine formulations (Friedlander and Little, 2009).

2.4 *B. ANTHRACIS* AS A BIOLOGICAL WEAPON

Research on *B. anthracis* as a biological weapon began more than 90 years ago, according to Inglesby and colleagues (2002). When appropriately prepared, *B. anthracis* spores are premier agents for biological warfare and bioterrorism because they can be produced in prodigious quantities in a form that is readily aerosolized and inhaled. These factors, plus the high mortality associated with inhalational anthrax, make *B. anthracis* a serious military and bioterrorism threat.

Most offensive biological weapons programs were terminated following the ratification or signing of the Biological Weapons Convention in the early 1970s (Inglesby et al., 2002). The United States' offensive biological weapons program was terminated before that by President Nixon in 1969. Nevertheless, international state programs to produce and weaponize *B. anthracis* have remained a concern. For example, in 1995 Iraq acknowledged having an anthrax weapons program to the United Nations.

Estimates of fatalities likely to occur after a major attack on an urban area using aerosolized *B. anthracis* as a bioweapon range into the millions (Meselson et al., 1994; Inglesby et al., 2002). Even an accidental release of aerosolized *B. anthracis* at a former Soviet military facility in Sverdlovsk in 1979 resulted in scores of infections and many human and animal deaths (see Box 2-1) (Leffel and Pitt, 2006). No inhalational anthrax fatality had occurred in the United States as the result of an act of war or terror until the 2001 anthrax mailings (Bush et al., 2001).

The events of 2001, when envelopes with *B. anthracis* spores were distributed through the U.S. mail, showed the potential of this microbe as a biological weapon. The resulting 22 cases of anthrax, including five fatalities, spread great fear and resulted in tremendous disruption and dislocation as mail distribution centers, congressional office buildings, and other sites suffered extensive

BOX 2-1
The Sverdlovsk Outbreak

In 1980, reports appeared in the international press of a widespread outbreak of anthrax in the city of Sverdlovsk (now Ekaterinburg) in the Soviet Union (Gwertzman, 1980). Soviet medical, veterinary, and legal publications reported that an outbreak had taken place in early 1979, involving livestock and humans who ate contaminated meat. In 1988, three Soviet physicians visited the United States to give formal and informal presentations on the outbreak (Meselson et al., 1994). In 1990, more discussion appeared in the Russian press (references in Meselson et al., 1994), leading Soviet President Boris Yeltsin to call for further investigations. Finally, in 1992, Yeltsin was quoted as saying "that the KGB admitted that our military developments were the cause" (references in Meselson et al., 1994).

A group of scientists from the United States carried out an on-site investigation during two visits in 1992 and 1993; their results were published in *Science* in 1994 (Meselson et al., 1994; Guillemin, 1999). They concluded that "the outbreak resulted from the windborne spread of an aerosol of anthrax pathogen, that the source was at the military microbiology facility, and that the escape of the pathogen occurred during the day on Monday, April 2 [1979]."

The outbreak has been used to model aerosolized agent release and parameters such as plume migration (Hogan et al., 2007), infectious dose (Brookmeyer et al., 2001; Wilkening 2006), incubation period (Brookmeyer et al., 2005; Wilkening, 2008), and the impact of public health intervention (Brookmeyer et al., 2001). A 1998 study by Paul Keim and Paul Jackson (Jackson et al., 1998) used the polymerase chain reaction (PCR) technique to analyze tissue samples from 11 patients from the 1979 Sverdlovsk outbreak. The study found that sequences representing the entire virulence complement (toxins and capsule) were present in all the samples, and also found that strains from at least four of the five known strain categories were distributed among the samples, as determined by variable region sequencing (variable number tandem repeats, or VNTRs). Two subsequent studies were published: one found additional evidence for multiple strains (Price et al., 1999) and the other found no evidence for multiple strains (Okinaka et al., 2008). Whether or not multiple strains were involved in this outbreak and if so the implications of this are not known. A previous analysis of 198 clinical samples from naturally occurring anthrax cases indicated the presence of only one strain in each case (Jackson et al., 1997).

contamination (see Chapter 3). Over 30,000 people were given prophylactic antibiotic treatment and six buildings required decontamination, all at a cost of over $1 billion (CDC, 2001a).

2.5 PHYLOGENY OF *B. ANTHRACIS*

B. anthracis is a member of the large genus *Bacillus* that includes other common and diverse species, such as *B. cereus, B. subtilis,* and *B. thuringiensis.*

Within the genus, *B. anthracis* represents a separate lineage that apparently evolved from the *B. cereus* parent species. Some strains of both *B. cereus* and *B. thuringiensis* are clearly closely related to *B. anthracis* (Kolsto et al., 2009). The production of endospores that resist chemicals, heat, UV light, and desiccation when the cells experience starvation or other environmental stresses is a common feature of members of the *Bacillus* genus.

B. anthracis has often been viewed as a highly monomorphic species, that is, one that shows little genetic variation among isolates. While it is the case that many bacterial species harbor much more extensive genetic diversity, newer molecular methods have allowed researchers to find genetic differences among natural isolates of *B. anthracis*, as discussed below. Moreover, all populations of bacteria are continually generating new mutations. While most new mutations remain very rare in a population, some of them may rise to high and measurable abundances, especially when they provide an advantage to the bacteria in terms of increasing their growth or survival under certain conditions. As discussed in detail in Chapters 5 and 6, the presence of new mutations among the spores in the attack letters provided an important lead in the anthrax letters investigation.

Modern scientific tools now afford an understanding at the molecular level of the similarities and differences among members of the genus *Bacillus*. For example, as noted above, the pathogenicity of *B. anthracis* is related to the presence of the pXO1 and pXO2 plasmids. These were once believed to separate *B. anthracis* from the other members of the genus, *Bacillus*. Within the last five years, however, *B. cereus* strains containing pXO1 or both pXO1 and pXO2 have been discovered, the latter in great apes in Africa. Thus, the presence of these plasmids is no longer believed to be the major factor separating *B. anthracis* from other members of the genus *Bacillus*. Rather, a specific mutation in *B. anthracis'* global regulator *plcR* gene, which controls the transcription of virulence factors in *B. cereus* and *B. thuringiensis,* now appears to be the key difference. All *B. anthracis* strains investigated carry a mutation that makes the *plcR* gene dysfunctional. This feature makes *B. anthracis* easily distinguishable from its close relatives.

Until the advent of modern molecular approaches, the genetic homogeneity of *B. anthracis* impeded efforts to reconstruct the evolutionary history of the species (Keim et al., 1999). Substantial progress has recently been made. Van Ert and colleagues (2007a) conducted a large study of 1,033 *B. anthracis* isolates using canonical SNP (single nucleotide polymorphism) analysis. Based on their results, they divided all *B. anthracis* isolates into A, B, and C lineages that had been previously recognized and further subdivided these into 12 distinct sublineages or subgroups (see Figure 2-1). Sublineage designation was based on whole-genome sequencing of seven isolates that represent major branches on a phylogenetic tree. The five other subgroups, whose genomes were not sequenced, appear at intermediate positions along these branches.

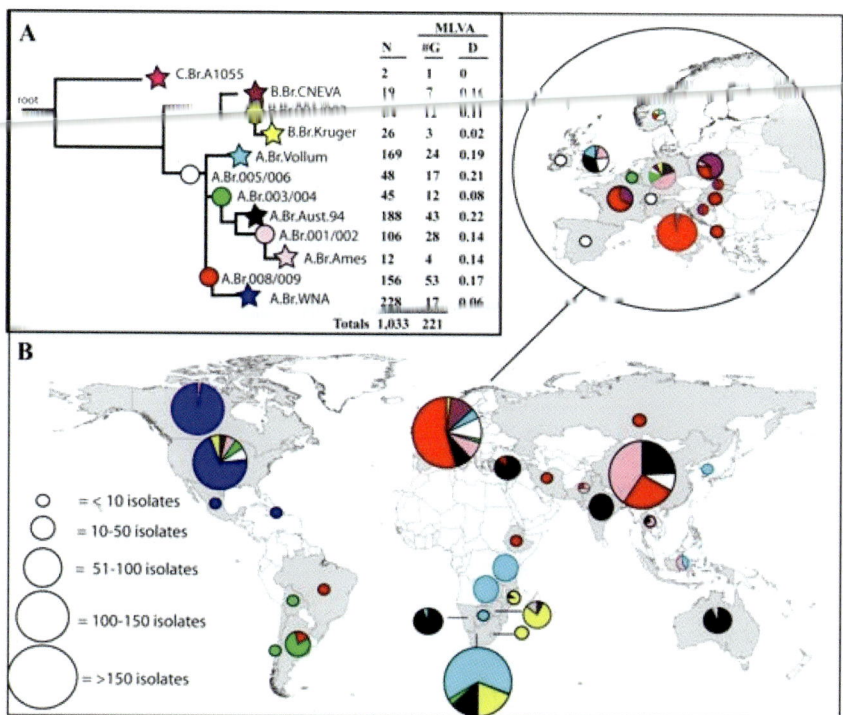

FIGURE 2-1 Worldwide Distribution and Lineages of *B. anthracis*.
The stars in this dendrogram represent specific lineages defined by one of seven sequenced genomes of *B. anthracis*. The circles represent branch points along the lineages that contain specific subgroups of isolates (Van Ert et al., 2007a).

Van Ert and colleagues (2007a) point out that dispersal of spores via commodities has distributed anthrax worldwide such that there are foci of the disease on all continents except Antarctica (see Figure 2-1). The more numerous "A" lineage isolates of *B. anthracis* are the most widely distributed, while there are more restricted distributions of "B" and "C" lineage isolates; for example, the B lineage is found mainly in South Africa and portions of Europe. There are also distinctive genotypes in the Western Hemisphere, with particular North and South American genotype subgroups that are rarely observed outside these regions. Based on molecular clock estimates, Van Ert and colleagues (2007a) note that the radiation of the A lineage seems to have coincided with periods of increased animal domestication and expansion of domestic animal populations. Although North America has a single dominant genetic type, in more recent times, trade in wool, skins, bone meal, and other products appears to have contributed to the introduction of an assortment of rarer *B. anthracis* genotypes on the North American continent. One of these, the "Ames" strain, is the strain found in the 2001 anthrax mailings.

2.6 THE EARLY HISTORY OF THE AMES STRAIN OF *B. ANTHRACIS*

The *B. anthracis* Ames strain was first isolated from a dead cow in Sarita, Texas, in 1981. Texas A&M University shipped this new isolate to the U.S. Army Medical Research Institute of Infectious Diseases (USAMRIID) at Fort Detrick in Frederick, Maryland. Because the box used for this shipment bore an old Ames, Iowa, address, the strain came to be known as the "Ames" strain. As this is a misnomer, at least in terms of the location of the origin of the first sample of the strain, Ravel and colleagues (2009a,b) called this specific isolate from Texas "*B. anthracis* Ames Ancestor." Chapter 5 describes the experiments used to determine that the Ames strain was the source of the material in the anthrax mailings.

The Ames strain is uncommon in nature. In the large study by Van Ert and colleagues (2007a), North America was represented by 273 isolates of *B. anthracis* spanning 44 genotypes but the Ames genotype was found only once, indicating that it is rare. According to Keim (2009), the Ames strain appears to be a recent immigrant to North America, with its closest relatives found in China. In North America, the natural distribution of the Ames strain appears to be limited to a small area of Texas. However, the Ames strain has been widely distributed as a laboratory strain. This fact, coupled with its rarity in nature, "makes it unlikely that the source material utilized in the 2001 bioterrorist attack was acquired directly from nature" (Van Ert et al., 2007).

2.7 SUMMARY

The bacterium *B. anthracis* is the causative agent of the disease anthrax. The infectious agent in all forms of the disease is the dormant *B. anthracis* spore. The spore's structural characteristics and properties provide the organism with resistance to environmental insults (e.g., dehydration, elevated temperature, UV irradiation) and also enable it to avoid the immune response of an infected person. Spores can remain viable in this dormant state for decades or more. Upon arrival in a hospitable environment (i.e., in a human or animal body) the spores can germinate and resume rapid growth within hours, causing illness.

In humans, the course and outcome of the disease depend on dose and on whether the infection is acquired via the skin, gastrointestinal tract, or inhalation. Most human cases involve skin lesions (cutaneous anthrax), whereby the infection is acquired as a result of handling infected animal hides or wool. Ingestion of food contaminated with *B. anthracis* can produce gastrointestinal anthrax, a much more serious disease. But the most lethal manifestation of anthrax is caused by inhalation of spores, causing a severe disease that is often fatal. Inhalational anthrax is generally rare and is mainly observed in individuals who work with animal skins. Inhalational anthrax in individuals not likely to have suffered occupational exposure, however, can be a sign of a biological attack with *B. anthracis* spores.

When appropriately prepared, anthrax spores are premier agents for biological warfare and bioterrorism because they can be produced in prodigious quantities in a form that can be aerosolized and inhaled. These factors, plus the high mortality associated with inhalational anthrax, make *B. anthracis* a serious military and bioterrorism threat. For these reasons, the appearance of *B. anthracis* in the 2001 mailings launched a major public health and criminal investigation.

3

Scientific Investigation in a Law Enforcement Case and Description and Timeline of the FBI Scientific Investigation

3.1 INTRODUCTION

The discoveries that individuals had contracted anthrax and that letters containing *B. anthracis* had been sent by U.S. mail launched a full-scale investigation by the U.S. Centers for Disease Control and Prevention (CDC), the U.S. Postal Inspection Service (USPIS), and the Federal Bureau of Investigation (FBI). Numerous investigative techniques were applied throughout the investigation, as outlined in Table 3-1. This chapter describes in brief the early stages of the investigation, specifically the gathering of evidence, formation of investigative teams, and decisions regarding scientific analyses that led to the FBI findings and conclusions that are summarized and evaluated more fully in Chapters 4 through 6.[1] It also introduces the concepts of science, scientific investigation in law enforcement, and the different views of uncertainty—and the manner in which it is described—in science versus law.

3.2 SCIENCE AND SCIENTIFIC INVESTIGATION AS PART OF A LAW ENFORCEMENT INVESTIGATION

In a scientific study, explanations for observable phenomena are sought through the gathering of reliable data and the formulation of testable hypotheses. Scientific observations must be reproducible and scientific hypotheses must be refutable. Science is typically an iterative, collaborative, and open process requiring the ability to pose hypotheses, test them, and pose new questions based on the resulting information. At times, a scientific investigation is a divergent process, in which new results drive research in several directions only

---

[1] Throughout this report, the committee describes its evaluation of the primary reports and data contained in the materials provided to the committee by the FBI. In cases where the committee was not able to review primary data, the committee's assessment of statements or analyses of data by others is provided.

TABLE 3-1 Timeline of Scientific Events in the Anthrax Mailings Investigation

Project initiated	Final report	Agency/ institution/ individual conducting the work	Project	FBI document number
10/4/01			First case of anthrax reported to CDC. The first lab confirmation of an initial clinical identification of *B. anthracis* from a victim of the letter attacks (Stevens) was done at the Florida State Laboratory in Jacksonville.	N/A
10/12/01	11/14/01	FBI and local law enforcement	Collection of biological evidence: 4 envelopes, 17 clinical samples, 106 environmental samples along mail paths (FL, DC, NJ, NY, CT)	N/A
10/17/01	10/19/01	Battelle Memorial Institute (BMI)	Microbiological analyses of letter material identifies 2 *Bacillus* species: one non-beta-hemolytic (consistent with *B. anthracis*) and one beta-hemolytic (not further characterized)	B2M1D1 B2M13D4
10/01		Beecher	Environmental sampling of mail bags	
10/01	11/01	CDC	Clinical isolates from stricken patients identified as *B. anthracis*	N/A
10/18/01	11/27/01	USAMRIID	Initial characterization of letter material (CFU, EM, visual inspection)	B1M1D2
10/01	11/26/01	Battelle Memorial Institute	SEM-EDX analysis of letter material	B2M13D3, B2M13D8
11/01		Armed Forces Institute of Pathology (AFIP)	SEM-EDX analysis of letter material	AFIP, 2001
10/01	05/02	Los Alamos National Laboratory (LANL)	Material analysis for evidence of genetic engineering	B1M4
10/01	9/02	Northern Arizona University	Identification of USAMRIID samples as Ames strain	B1M3

TABLE 3-1 Continued

Project initiated	Final report	Agency/ institution/ individual conducting the work	Project	FBI document number
Fall 2001	2/28/02	BMI	Particle size distribution performed on letter samples and some surrogate samples, but not the Dugway Proving Ground surrogates	B2M13D11
Fall 2001		CDC	Combined epidemiological analysis of all case related to the anthrax mailings	
11/07/01	12/03/01	CDC	Using phenotypic substrate utilization and 16S rDNA sequencing, *Bacillus* contaminant identified as *B. subtilis*	B2M1D2
11/09/01	11/09/01	FBI	SEM analysis of envelopes	B1M7D16
Fall 2001			Consortium of agencies (NSF, NIH, DOE, DOJ, FBI, USDA, DOD, and Intelligence Community) formed to advise investigation and provide support resources	N/A
Fall 2001		BMI	Silicon incorporation into spore coat	B2M13D7
Fall 2001	Spring 2002	The Institute for Genomic Research (TIGR)	Completion of genome sequence for *B. anthracis* Porton and "Ames 2001 Florida strain" (clinical isolate); publication in *Science*, June 2002	B1M5D1, B1M5D3
Fall 2001	Spring 2002	USAMRIID	Detection of phenotypic variants ("morphotypes") among colonies derived from letter spores	B1M2D12
12/01	12/06	Dugway Proving Ground (DPG)	Reverse engineering of spore "powders"	B1M13 B1M14

continued

TABLE 3-1 Continued

Project initiated	Final report	Agency/ institution/ individual conducting the work	Project	FBI document number
2/02	10/14/02	Lawrence Livermore National Laboratory (LLNL), National Ocean Sciences Accelerator Mass Spectrometry Facility National Ocean Sciences Accelerator Mass Spectrometry Facility (NOSAMS), Woods Hole Oceanographic Institute	Carbon dating by Accelerator Mass Spectrometry	B1M8
2/02		FBI	Subpoenas of laboratories for samples of *B. anthracis* Ames	
2/02	10/06	Sandia National Laboratory (SNL)	Elemental analyses (SEM-EDX) of letter material, envelopes, and DPG surrogates (final report not dated)	B1M7
2/02	10/06	SNL	Silicon analyses	B1M6
2/01/02	8/13/05	University of Maryland (UMD)	Agar analysis	B1M10
3/07/02	2/01/06	DPG	Analytical chemistry analysis of spore powders	B1M13
3/22/02	4/14/02	FBI	Volatile organic compound analysis in evidentiary material	B1M7D2
Early 2002	7/6/05	FBI	ICP-OES: elemental composition of letter material, culture media, envelope types	B1M6 B1M7
12/02	6/1/04	TIGR	Whole genome sequencing of Morphs A, B, C, D	B1M5
8/02	2/22/04	UMD	Heme analysis	B1M10
8/02	8/05	*Edgewood Chemical Biological Center (Army)* ECBC	Agar and heme analysis	B1M10
8/02	9/02	BMI	Agar and heme analysis	B1M10

TABLE 3-1 Continued

Project initiated	Final report	Agency/ institution/ individual conducting the work	Project	FBI document number
10/02	2/04	Commonwealth Biotechnologies, Inc. (CBI)	Contract to develop Morph A assays; only A1 and A3 were validated	B2M5
3/03	5/05	University of Utah	Stable isotope signatures	D1M9
3/14/03	10/10/03	Applied Biosystems (AB)	16S rDNA sequencing of Brokaw letter *B. subtilis* contaminants	B2M1D4
7/31/03	8/11/03	Novozymes Biotech, Inc.	*B. subtilis* contaminant compared to *B. licheniformis*	B2M1D3
10/03	5/14/06	TIGR	Multiple locus PCR-based assay for direct comparison of *B. subtilis* strains to Post *B. subtilis*	B1M5D2
10/03	6/4/05	TIGR	Whole genome sequencing of Morph E and *B. subtilis* contaminant (Post and Leahy)	B1M5
10/10/03	6/30/05	AB	Genome sequencing of *B. subtilis* "H2122" (not identified elsewhere)	
3/04	2/06	CBI	Repository (1104 samples) screening for A1 and A3; second screening (300 samples) 7/07-10/07)	B2M5D8
6/25/04	7/05	CBI	Contract to develop assays for Morphs B and D: both rejected	B2M6
7/13/04	9/14/04	USAMRIID	Screening of selected samples of FBIR for presence of Morphotypes	B1M2D13
7/23/04	6/7/05	Midwest Research Institute (MRI)	Contract to develop assay for Morphs B and D; Morph B assay rejected; Morph D assay accepted	B2M8
11/19/04	4/05	IIT Research Institute (IITRI)	Contract to develop assays for Morphs B and D; Morph B assay rejected; Morph D assay accepted	B2M7
12/01/05	10/10/07	MRI	Repository screening for Morph D	B2M8

continued

TABLE 3-1 Continued

Project initiated	Final report	Agency/ institution/ individual conducting the work	Project	FBI document number
5/05	4/24/07	IITRI	Repository screening for Morph D	B2M7
12/05	1/07	TIGR	Contract to develop assay for Morph E	B2M9
10/06	7/07	Pacific Northwest National Laboratory (PNNL)	Agar and heme analysis	B1M11
11/06	12/07	NBFAC	Repository screening for *B. subtilis* contaminant	B2M4D2
2/15/07	12/04/07	CBSU (FBI) National Bioforensic Analysis Center (NBFAC)	*B. subtilis* analysis by real-time PCR: screening of repository and other samples	B2M4
6/25/07	8/25/07	TIGR	Repository screening for Morph E	B2M9
8/04/07	11/30/07	National Center for Agricultural Utilization Research (NCAUR)	Genetic diversity and phylogenetic analysis of *B. subtilis* samples	B2M3
8/29/07	8/29/08	FBI	Analysis of meglumine and diatrizoate in RMR-1029, letter material, other samples	B1M12
10/05/07	10/08/08	TIGR	Finalization of *B. subtilis* genome sequence	B2M2D2
1/09/08	6/08/08	TIGR	Screening of unidentified *B. subtilis* isolates for presence of sequence specific to *Post/Brokaw* contaminant	B2M2D3, B2M2D4
3/27/08	9/30/08	University of Cincinnati	Statistical analysis of FBIR screening data	B2M10

to converge again when more information informs decisions about which directions to pursue. This divergence and convergence make a scientific investigation different from a law enforcement investigation, in which the drive toward convergence dominates to a greater degree. In addition, the approach used to gather, process, and analyze evidence can differ between a purely scientific investigation and a law enforcement investigation. Both types of approaches are necessary in a bioterrorism investigation, which also requires attention to public health risks and safety needs (see Box 3-1).

An important feature of science is that observations are made in a manner that is independent of the observer and on the assumption that other observers can and would make the same observations. Science relies on validated methods for gathering observations and making quantitative measurements systematically and reproducibly. Standards must be set for collecting data under controlled and well-specified conditions, assessing possible sources of error, establishing causality (and acknowledging that a relationship is only a correlation when causality cannot be inferred and supported), and applying empirical findings to validate or refute particular hypotheses. New scientific methods must be assessed for their accuracy and reliability, their limitations, and the range of circumstances under which they can be appropriately applied.

The Qualifiers of Certainty in the Biological Sciences

A key question in this study was "Based on the available data, how strong is the apparent association between the letter evidentiary material and a particular source or sample (e.g., flask RMR-1029)?" Some of the committee's most important findings focus on the strength of a given association and on the conclusions that one should draw from the available scientific data about the nature of the association. Thus, it is important to review briefly the use of terminology to describe the strength of an association.

Quantifying an association, as well as the degree of certainty (or uncertainty) in that association, involves statistical methods (see Chapter 6). Common language involves *qualifiers*, rather than *quantifiable* measures, of this association and the degree of confidence in it, which can cause confusion among practitioners from different fields that use the terms. Since the interpretation of these qualifiers and the ways in which they are used differ across disciplines (e.g., statistics, science, law, common language), their use by the committee is clarified here. In the chapters that follow, the committee uses the following four qualifiers of association, listed in order of increasing certainty (decreasing uncertainty):

- consistent with an association
- suggest an association
- indicate an association
- demonstrate an association

BOX 3-1
Bioterrorism Investigations

A crime scene typically is a place where the victims and perpetrators meet in time and space. The traditional crime scene, which may have multiple locations, is the logical place to search for physical evidence leading to the identity of the perpetrator(s). Identifying, collecting, and preserving probative evidence combined with investigative detective work is the usual approach to a successful prosecution. Solving the crime is the ultimate goal of the scene investigation, but there also are other reasons to investigate the crime scene, including: 1) developing investigative leads for detectives; 2) developing specific information in the form of evidence or investigative logic to enable a successful prosecution; 3) locating, collecting, and preserving probative physical evidence that can provide evidence of innocence or guilt; 4) developing information and physical evidence that provides an accurate reconstruction of the events of the crime; and 5) linking multiple crimes through the evidence collected across sites (USDOJ, 2000; Fisher, 2005).

All crime scene investigations require the integration of multiple forensic disciplines through the juxtaposition of science and scene investigative skills. Scientific criminal investigations require an amalgamation of capabilities including scene experience, attention to detail, a skeptical perspective, powers of observation, and the application of logic (Gardner, 2005).

Although bioterrorism event scenes have elements in common with other crime scenes, such as the identification, collection, and preservation of the forensic evidence, they can also differ because of the inherent risks to investigators and to the public. Also, unlike traditional crimes, they may not always involve a location where the participants meet in time and space, as shown by the 2001 *B. anthracis* mailings in which the dissemination of *B. anthracis* occurred by means of the United States Postal Service (Jernigan et al., 2002). The perpetrator(s) presumably worked at a distance, so the criminal investigation spanned several locations, including those of the envelopes, post offices, and street postal boxes that might have held contaminated envelopes, and the location(s) where the *B. anthracis* might have been manufactured. Each scene required a comprehensive and coordinated investigation to find, collect, and preserve the *B. anthracis* spores.

Future bioterrorism events may differ in the nature of the biological agent or toxin and in the mode of delivery. In general, bioterrorism incidents can be expected to be handled differently than the typical homicide scene investigation because such events require both traditional scene management skills and the special requirements of scenes involving bioagents. Bioterrorism investigators must consider issues such as public safety, operational planning, sampling strategy, packaging, transport, and storage. The immediate imperative to consider public health needs requires finding and collecting the biological agent so that it can be identified expeditiously, as well as defining and containing the environmental risk to those not yet exposed. The investigation and containment must be accomplished while ensuring the safety of investigators and the public during the investigation and while remediating the scene. Sampling strategies must combine the collection and preservation of bioagents with the collection of usual forensic evidence (Budowle, 2006). In some future scenarios, delivery of the biological agent might occur through natural routes of biological transmission, and thus the "crime scene" may be limited to the site at which the biological agent was prepared or delivered.

The expression "consistent with" is frequently used in this report and conveys the weakest level of certainty (greatest amount of uncertainty). In general, when the term "consistent with" is used, it means that an association may or may not be present; the available data can neither rule out nor confirm an association. The term "suggests" denotes a greater level of certainty for an association than "consistent with," but even here the normal use of the word in science denotes a weaker level of certainty than is implied by the word in everyday parlance. That is, the potential for an association is stronger, and the evidence for the absence of an association is weaker, but both are still possible. In contrast, the terms "indicate" and "demonstrate" denote higher degrees of certainty and these are usually reserved for strong scientific conclusions (i.e., less uncertainty, or less likelihood of an absence of an association). All four levels could potentially be quantified with measures of "statistical significance," but the committee does not assign such measures in most instances because the data at hand are generally not appropriate for such precise quantification of the degree of uncertainty.

In summary, the reader is cautioned to consider carefully the terminology in this report in light of the fact that the qualifiers of certainty used here are those used most commonly in the scientific literature and that these words can carry different weight in common language and in the courtroom.

3.3 THE FEDERAL COORDINATED RESPONSE AND ASSIGNMENT OF LABORATORY WORK

Oversight and coordination of a complex scientific study are critical. Large teams of scientists have been successful at complex studies (e.g., in particle physics and genomic research) because they have a clear leadership structure for the coordination and planning of their efforts (International Human Genome Sequencing Consortium, 2001; Venter et al. 2001; NRC, 2003). With pressures of time and expense in any study, someone or some group in the research team must make decisions about which avenues to pursue and which to abandon.

In 2001, the FBI had a science laboratory at Quantico, the Hazardous Materials Response Unit, and another team of weapons of mass destruction (WMD) experts, but it did not have the capabilities to handle all of the types of scientific experiments that would be required to examine the evidence in the investigation of the *B. anthracis* mailings. FBI investigators quickly realized the need to turn to outside laboratories and experts for help. The Bureau immediately formed an internal group that had members from the scientific team, investigative team, and terrorism team. Recognizing the importance of parallel criminal and scientific investigations, the FBI embedded high-level DOJ staff from the internal team, who then advised them throughout the investigation.

At that time, the FBI did not have the organizational structure needed to oversee such a complex, multifaceted, and involved scientific investigation.

In 2003, the agency remedied this organizational limitation by forming a new unit focused on investigations involving chemical, biological, radiological, and nuclear sciences, called the Chemical, Biological, Radiological, and Nuclear (CBRN) Sciences Unit, or the Chemical Biological Science Unit (CBSU).

The committee was told that the team working on the scientific investigation met weekly with the law enforcement team for information sharing, strategy, and coordination. Reports and notes from some of these meetings were shared with the committee late in the process of finalizing this report. The reports indicate the complexity of the parallel tracks of the investigation and document progress of each aspect of the scientific efforts and the decisions to proceed with or abandon particular lines of the investigation.

In the early stages of the investigation, the FBI sought the advice of outside experts to assist in characterizing the properties of the *B. anthracis* evidence. The engagement of these experts was aided by the creation of an advisory group led by the director of the National Science Foundation, the director of the National Institute of Allergy and Infectious Diseases of the National Institutes of Health, and federal officials from numerous other science agencies. This group met regularly in classified sessions with FBI leaders to hear about the investigation and to provide advice and the names of potential subject matter experts the FBI could engage for assistance.

In addition, several Technical Review Panels were formed consisting of scientists from Department of Energy National Laboratories, academic laboratories, and members of the National Academy of Sciences. Panels were constituted to review the analytical plan (that is, what tests should be done and by whom), the progress of the investigation, and the chemistry and biology techniques used. We reviewed several reports from meetings held in late 2001 and thereafter (FBI Documents, B1M1, B3D1-7). According to these materials and the DOJ report, "At the outset of the investigation, three panels comprised of 33 of the nation's leading authorities in bioweapons development from the former offensive bioweapons program, microbiology, chemistry, and microscopy were convened to assist the FBI in developing a comprehensive analytical framework to evaluate the anthrax powders recovered from the envelopes and the contamination found in the AMI Building" (USDOJ, 2010, p. 13).

The FBI benefitted from these early informal and regular meetings of senior leadership from other science agencies (FBI, 2009; Colwell, 2009). The FBI also received input from the Department of Defense (DOD), the Intelligence Community, DOJ, CDC, and Armed Forces Institute of Pathology (AFIP) regarding the scientific investigation (FBI/USDOJ, 2011).

As is shown in Table 3-1, the U.S. Army Medical Research Institute of Infectious Diseases (USAMRIID) played a central role in the scientific investigation. The facility had provided analytical services to the FBI Laboratory since 1998. During October and November 2001, scientists at USAMRIID were included in the team performing on-site testing at the American Media,

Inc. (AMI) building in Florida, and they conducted the initial examinations of the letter spore preparations for physical characteristics (using microscopy and electron microscopy) and spore viability (see Chapters 4 and 5). USAMRIID scientists also conducted microbiological analyses and identified the dominant and variant morphological colony types that appeared in the evidentiary material (see Chapter 5).

The FBI also sought the help of dozens of outside laboratories. In total, thousands of samples were processed and analyzed by 29 academic, government, and private BSL-3 laboratories across the country (Piggee, 2008; USDOJ, 2010). The work commissioned by the FBI was highly compartmentalized. Most of the laboratories conducting analyses were not aware of other analyses under way (Keim, 2009; Michael, 2009; Weber, 2009). Scientists at these laboratories responded rapidly and provided the bulk of the scientific studies on which the FBI relied in its investigation. The committee read reports prepared by outside scientists responding to specific requests from the FBI and we received reports of periodic reviews of contracted work by panels of experts (B3D1-7). The committee also reviewed reports of work carried out in parallel at the AFIP although it is not clear how closely AFIP and the FBI investigative and scientific teams worked together or coordinated their efforts.

As the scientific investigation proceeded, several laboratories conducted sequential and parallel scientific analyses on the evidentiary material gathered from the letters, environmental samples, and clinical samples (see Table 3-2). These analyses first focused on identifying the nature of the letter and environmental materials, their similarities and differences, their biological, chemical, and physical properties, and, eventually, their similarity to other samples of Ames strain *B. anthracis* in laboratories around the world. According to the affidavit in support of a search warrant submitted by Postal Inspector Thomas F. Dellafera (Case number O7-524-M-01, October 31, 2007), 16 domestic laboratories and three foreign laboratories (in Canada, Sweden, and the United Kingdom) were identified as having the Ames strain in their inventories prior to the attacks; two additional domestic laboratories were subjected to consent searches, and one domestic laboratory was subjected to a search warrant. A subpoena prepared by the FBI in early 2002 for sample submission specified that only Ames strain samples be submitted; the subpoena protocol for these submissions is described in Chapter 6. In the end, the FBI assembled a repository of over 1,070 Ames strain samples from 20 laboratories, of which 1,059 were viable. Attributes of the samples in the repository were compared against the characteristics of the evidentiary samples (as discussed in Chapters 5 and 6).

According to FBI officials, the focus of these analyses was to provide FBI investigators with scientific leads that could be used to assist in its criminal investigation (FBI, 2009). Chapters 4 through 6 provide in-depth descriptions of the analyses conducted, conclusions reached, and this committee's findings and lessons learned for the future.

TABLE 3-2 Analytical Techniques Used on the Evidentiary Material

Technique	Florida environmental and clinical samples	DC, NY, CT, NJ clinical samples	New York Post letter	Brokaw letter	Daschle letter	Leahy letter
MLVA (genetic analysis of strain)	X (B1M3D1)	X	X	X	X	X
Sequencing of *pagA* genes	X (B1M4D2)	X	X		X	?
Complete genome sequencing of wild-type *B. anthracis*	X		X			X
Sequencing and PCR analysis to determine whether genetic engineering occurred	X		X	X	X	?
Partial genome sequencing of *B. subtilis*			X	X	No *B. subtilis*	No *B. subtilis*
Whole genome sequencing of *B. subtilis*			X	Not sequenced	No *B. subtilis*	No *B. subtilis*
Phenotypic screens of variants (morphotypes)	X (B1M2D14)		X		X	X
Whole genome sequencing of morphotype A			X			X
Whole genome sequencing of morphotype B			X			X
Whole genome sequencing of morphotype C						X
Whole genome sequencing of morphotype D						X
Whole genome sequencing of morphotype E					X	X

TABLE 3-2 Continued

Technique	Florida environmental and clinical samples	DC, NY, CT, NJ clinical samples	New York Post letter	Brokaw letter	Daschle letter	Leahy letter
SEM analysis to assess size and shape of spores			X		X	X
Assessment of bulk silicon content			X			X
SEM-EDX bulk analysis of silicon			X		X	X
SEM-EDX to assess silicon in spore coat			X		X	X
SEM-EDX bulk elemental analysis			X		X	X
SEM-EDX for spatial resolution and sensitivity			X			X
Assays for presence of agar			X			X
Assays for presence of meglumine diatrizoate			X			X
Headspace GC-MS and infrared spectroscopy to detect VOCs						X
Radiocarbon analysis for dating						X
Stable isotope analysis to determine growth medium						X
Envelope analysis			X		X	X
Assays of genotypic variants (A, B, C/D, E)			X		X	

GC-MS = gas chromatography-mass spectroscopy; MLVA = multiple-locus VNTR [variable-number tandem repeat] analysis; PCR = polymerase chain reaction; SEM = scanning electron microscopy; SEM-EDX = scanning electron microscope with energy-dispersive X-ray analysis; VOC = volatile organic compounds

3.4 COLLECTION AND ANALYSIS OF CLINICAL AND ENVIRONMENTAL SAMPLES AND CROSS CONTAMINATION

Beginning in October 2001, investigators collected biological evidence from a variety of sources and locations. By the end of the collection phase, *B. anthracis* isolates were gathered from 4 powder-containing envelopes, 17 clinical specimens obtained from infected patients, and 106 locations along the mail path of the implicated envelopes from Florida, the National Capital Region (Washington, D.C.), New Jersey, New York, and Connecticut, and overseas (Jernigan, 2002).

3.4.1 Clinical and Epidemiological Samples

As noted in Chapter 2, anthrax is generally a rare disease that most often occurs in people who have contact either with infected livestock or with contaminated animal skins or other animal products. Inhalational anthrax is particularly rare (Bush et al., 2001), and the few cases reported in the United States have been due to occupational exposures. Before 2001, the last case of inhalational anthrax in the United States was reported in 1976 (Jernigan et al., 2002). Thus, the occurrence of inhalational anthrax in Robert Stevens, an employee of AMI, was not immediately recognized as such.

Larry Bush, a physician with training in microbiology, was the first to surmise that the Gram-positive bacilli in the patient's cerebrospinal fluid (CSF) were *B. anthracis* (Cole, 2009). Phil Lee, at the Florida State Laboratory, provided confirmation based on various tests, including lysis by a *B.anthracis*-specific virus. Researchers in CDC laboratories used standard microbiological techniques to verify the putative agent as *B. anthracis* (Jernigan et al., 2002).

Bush and colleagues (2001) point out that in the Stevens case, although there was no typical "occupational exposure" to infected animals or contaminated animal products, the victim was apparently infected at work. No spore-containing letter was ever found, but coworkers reported that on September 19, 2001, Stevens had closely examined a suspicious letter containing powder. Anthrax spores were later found on the victim's computer keyboard, at other locations in AMI, in asymptomatic coworkers, in the AMI mailroom, and at regional and local postal centers that served the AMI worksite. Three days after his hospital admission on October 2, 2001, Robert Stevens died, becoming the index case for this 2001 bioterrorism event (Bush et al., 2001).

In the following week, as illustrated in Table 1-1 (Chapter 1), cases of cutaneous anthrax were reported in New York City, subsequently linked to exposure to letters containing suspicious powder (Greene et al., 2002). Ultimately, four envelopes containing *B. anthracis* spores were recovered from media outlets in New York and from government offices in Washington, D.C., or the containment facility created to store potentially contaminated mail. All four envelopes were postmarked by the United States Postal Service (USPS) Trenton Processing and Distribution Center in New Jersey. On October 18, a case of cutaneous anthrax was confirmed in a postal worker at the Trenton Center, which was then closed. An investigation by CDC and the New Jersey Department of Health was initiated (Greene et al., 2002). Subsequently, CDC and other agencies conducted a combined epidemiological analysis of all cases related to the anthrax mailings (Jernigan et al., 2002). This analysis was a coordinated effort among medical and laboratory facilities and local, state, and federal public health and law enforcement agencies. CDC's Emergency Operations Center supported local, state, and federal public health investigators in Florida, New York City, New Jersey, the District of Columbia, and Connecticut. Teams also coordinated with the USPIS, the Department of Defense, the FBI, and other federal organizations.

Jernigan and colleagues (2002) summarized the early overall epidemiologic findings. As illustrated in Figure 3-1 from the CDC, a total of 22 cases of bioterrorism-related anthrax were identified—11 inhalational cases and 11 cutaneous cases. Victims were identified in seven states along the east coast: Florida (2 cases), Maryland (3), New Jersey (5), New York (8, including one New Jersey resident exposed in New York), Connecticut (1), Pennsylvania (1), and Virginia (2). Five of the inhalational anthrax patients died, resulting in a fatality rate for inhalational anthrax of 45 percent. Eight of the 11 cases of inhalational anthrax were confirmed as *B. anthracis* from analysis of clinical specimens. In most cases, the exposures likely occurred in the victim's place of work. More than half the victims (12 of 22) were mail handlers at the USPS or in government and media industry mailrooms, eight of whom developed inhalational disease and four cutaneous disease. Of the remaining 10 victims, 6 were other media company employees working at locations where mail containing powder was presumed to have been received (AMI, CBS, NBC, and the *New York Post*). A seventh victim, a 7-month-old child had visited ABC's offices in New York.

Three other victims—a 61-year-old Manhattan hospital supply room worker, a 51-year-old bookkeeper from New Jersey, and a 94-year-old Connecticut resident—appeared to have no workplace associations. Since postal processing facilities were widely contaminated with *B. anthracis*, the FBI suggested that cross-contaminated mail was the source of the exposures for these three patients (Jernigan et al., 2002; Greene et al., 2002). Jernigan and colleagues (2002) noted that "The possibility of *B. anthracis* exposure from envelopes secondarily contaminated from implicated postal facilities greatly extended the group of potentially exposed persons in our investigation."

The epidemiological investigations found two distinct case clusters separated in time. The first case cluster was related to the two envelopes postmarked September 18, which were apparently mailed in or around Trenton, New Jersey, and delivered in New York City. As shown in Figure 3-1, 11 cases in all were classified as part of the cluster related to the September 18 mailings. Seven cases of cutaneous anthrax occurred in media company employees in New York (5) and postal workers in New Jersey (2). The two cases of inhalational anthrax

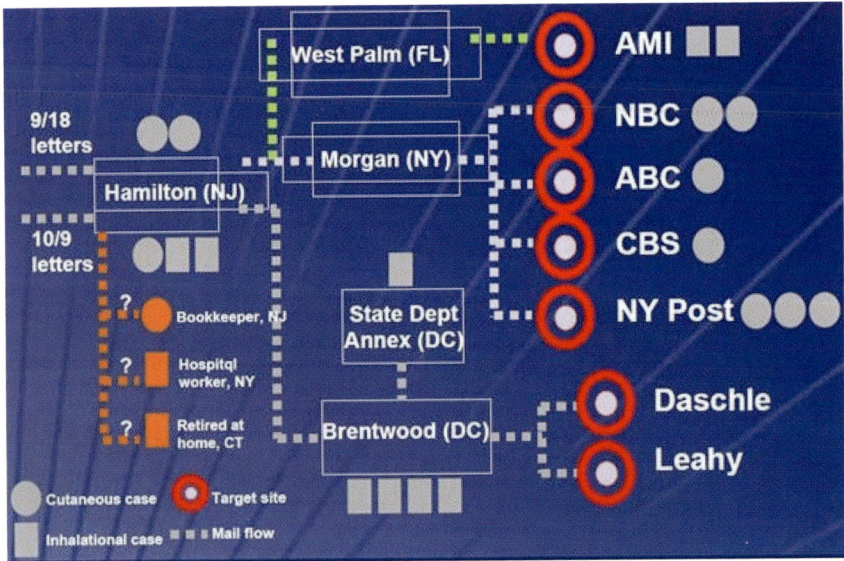

FIGURE 3-1 Trajectory and outcomes of anthrax mailings.
SOURCE: CDC. This image is a work of the Centers for Disease Control and Prevention, taken or made during the course of an employee's official duties. As a work of the U.S. federal government, the image is in the public domain.

in Florida were also assigned to this cluster, based on the hypothesis that an unrecovered mailing had been sent there. *B. anthracis* was isolated from environmental samples at six postal facilities along the path followed by the mail to AMI as well as at the AMI building itself. The dates of onset of illness in the two AMI employees in Florida were also consistent with exposure to envelopes mailed in mid-September. Victims in the cluster that followed the September 18 mailings were more likely to have cutaneous disease and to have been exposed at news media sites (Jernigan et al., 2002).

The second case cluster involved the two recovered envelopes sent to the offices of Senators Tom Daschle and Patrick Leahy, also mailed in or around Trenton and postmarked October 9, 2001. All five cases from the D.C. metropolitan area were part of this cluster and all five of these victims contracted inhalational anthrax and worked in postal facilities contaminated by the October 9 letters. Two additional cases of inhalational anthrax occurred in postal employees in New Jersey. In general, the October 9 mailings were associated with more severe illness. Victims in the second cluster were more likely to have been exposed at mail handling facilities along the path to the Senate offices.

Chapter 4 examines the differences in the physical appearance and consistency of the powders between the September and October mailings as reported

SCIENTIFIC INVESTIGATION IN A LAW ENFORCEMENT CASE   63

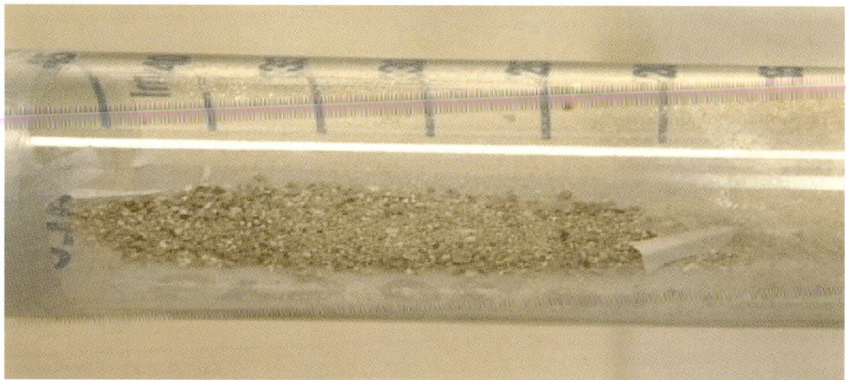

FIGURE 3-2 *New York Post* letter powder.
SOURCE: FBI. This image is a work of the Federal Bureau of Investigation, taken or made during the course of an employee's official duties. As a work of the U.S. federal government, the image is in the public domain.

FIGURE 3-3 Leahy letter powder.
SOURCE: FBI. This mage is a work of the Federal Bureau of Investigation, taken or made during the course of an employee's official duties. As a work of the U.S. federal government, the image is in the public domain.

by the FBI (see Figures 3-2 and 3-3). In brief, the powders from the *New York Post* and Brokaw letters were multicolored and granular in consistency. The envelopes in the second wave of mailings contained a powder of uniform color and smaller particle size, which would facilitate more efficient airborne transmission of the *B. anthracis* spores and could account for the prevalence of inhalation cases in the second cluster (Hassell, 2009). In addition, despite the fact that the envelopes in the October mailings were unopened during their passage through the postal centers, the use of high-speed processing and sorting machines may have contributed to dispersal of the spores and exposure of postal workers. Greene and colleagues (2002) reported that, although the envelopes containing the *B. anthracis* were handled in only a small area of the Trenton facility, environmental sampling found evidence of spores throughout the facility.

The *B. anthracis* isolates cultivated from the clinical specimens of patients, the four recovered powder-containing envelopes, and over 100 environmental samples collected along the suspected path traveled by the contaminated mail were subtyped by the CDC using multiple-locus variable-number tandem repeat analysis and sequencing of the protective antigen gene (*pagA*). In addition, *pagA* was amplified and sequenced directly from some clinical specimens. All of the results indicated the presence of the *B. anthracis* Ames strain (Hoffmaster et al., 2002).

Jernigan and colleagues (2002) noted that their epidemiological investigation had several limitations. Because identification of case patients involved numerous local, state, and federal officials, data collection methods were not uniform. The widespread use of postexposure prophylaxis and the difficulty of obtaining information about potentially exposed persons prevented accurate estimates of anthrax exposure rates. Some cases may have been overlooked because patients might have been administered antimicrobials after being mistakenly diagnosed with other types of infectious diseases, without physicians recognizing the disease to be anthrax. The lack of prior experience with bioterrorism also forced the investigators to refine methods and redefine interventions on a continuing basis.

3.4.2 Crime Scene Environmental Samples

The FBI hazardous materials (HAZMAT) team, with the assistance of scientists from USAMRIID, CDC, USPIS, Environmental Protection Agency (EPA), and contractors, performed environmental investigations to assess the presence and extent of *B. anthracis* contamination and to guide decontamination and environmental remediation (Jernigan et al., 2002). Environmental samples were collected at contaminated worksites and mailboxes by public health, law enforcement, and other government and contract staff (CDC, 2001a; Sanderson, 2001, 2002). At the time, the FBI did not have biosafety level 3 (BSL-3) laboratory facilities capable of handling *B. anthracis*, so samples were tested at laboratories participating in the local, state, and federal investigation efforts as described in Chapters 4 through 6.

Crime scene environmental samples were collected by surface sampling, mostly with RODAC (replicate organism detection and counting) contact plates. Swabs, wipes, high-efficiency particulate air (HEPA) vacuum filtration, and air sampling also were used (Dull et al., 2002; Jernigan et al., 2002; Teshale et al., 2002; Beecher, 2006). Based on the documents provided to the committee, it appears that the FBI and CDC relied on culture-based techniques to detect *B. anthracis* in the environmental samples (Jernigan et al., 2002; Hoffmaster et al., 2002). In much of the work on samples collected from the environment, colonies propagated on agar plates were presumptively identified as *B. anthracis* based on morphology (Beecher, 2006). Selected colonies were definitively identified as *B. anthracis* using standard confirmatory tests. At least one environmental swab sample from AMI was sent to Patricia Worsham at USAMRIID (in June 2005) for detection and identification of *B. anthracis* variant colony morphotypes. Material from this swab was used to inoculate sheep blood agar. The report of this work by Worsham (2009; FBI Documents, B1M2D14) states that *B. anthracis* variant morphotypes A, B, and C/D were found, but not morphotype E, in addition to the wild-type colony morphotype. Furthermore, Worsham states that a *Bacillus* strain was recovered that resembled the *B. subtilis* found in the *New York Post* letter. The report states that in October 2006, cell suspensions from 34 colonies that exhibited a variant morphotype, as well as from the *B. subtilis*-like isolate, were sent to the National Bioforensic Analysis Center (NBFAC) for DNA extraction, and that the DNA from the variant colony morphotypes were to be sent to the Institute for Genomic Research (TIGR) for sequencing of the morphotype A, B, C/D, and E genomic regions. However, according to statements by the FBI to this committee (FBI/USDOJ, 2011), the U.S. Attorney's Office advised that this sequencing and further characterization of these colony morphotypes from AMI would not be undertaken.

The FBI did not use molecular assays for detecting *B. anthracis* DNA directly in environmental samples (in the absence of a cultivated isolate). Molecular assays targeting *B. anthracis*-specific genetic markers had been developed prior to 2001 for detecting this organism in food (Yamada et al., 1999) and in environmental samples (Beyer et al., 1995). An on-site polymerase chain reaction (PCR)-based device was used by the CDC at the Brentwood postal facility in 2001 for preliminary assessment or adjunct analysis of *B. anthracis* DNA in environmental samples, but this approach had not been validated by them at that time for these types of samples. However, the Biological Aerosol Sentry and Information System was deployed in 2001 for environmental monitoring and incorporated PCR-based detection methods for a variety of biological agents (Fitch et al., 2003).

The committee recognizes that the FBI may have hesitated to apply a newly emerging method to the assessment of forensic evidence before it had been widely adopted or validated for this purpose; however, under exigent circumstances, it can be appropriate to proceed without completing full validation. The FBI stated to the committee (FBI/USDOJ, 2011) that laboratories in the CDC Laboratory Response Network that were equipped for PCR analysis were overwhelmed with samples early in the investigation, and that the FBI obtained more timely results on the presence of *B. anthracis* in environmental samples by relying instead on RODAC plate isolation techniques. In any case, by 2004, PCR had been validated for various bacterial agents in environmental samples (Malorny et al., 2004) and could have been performed on the environmental samples from 2001.

In addition to the environmental sampling, nasal swab specimens were collected from potentially exposed individuals to help delineate the area of exposure to aerosolized spores and to determine where persons with inhalational anthrax might have been exposed based on where they worked. According to Jernigan and colleagues (2002), because the sensitivity of nasal swab cultures diminishes with time following human exposure, attempts were made to obtain cultures within seven days of exposure. The presence of *B. anthracis* from nasal swab cultures was determined by Gram stain and colony characteristics as well as through confirmatory testing by laboratories participating with the local, state, and federal efforts (Jernigan et al., 2002).

Finally, in the new materials provided to the committee it is noted that PCR analysis was performed on human remains from United flight 93 on 9/11/2001 that were identified as those of the hijackers (B3D1). Analysis was performed at USAMRIID and at AFIP for sequences diagnostic of *B. anthracis*. One assay at USAMRIID gave positive results, but these results were believed by the FBI to be due to laboratory contamination. All other results were negative. As the committee learned at the January 2011 meeting, there were no tests done on remains from any of the other September 11, 2001 hijackers.

3.4.3 Samples from an Overseas Site Identified by Intelligence

In December 2010-January 2011, the FBI first made available to the Committee "AMX Weekly Science Updates" and a newly de-classified document that described the collection and analysis of environmental samples from an undisclosed site outside the continental United States (OCONUS) for the presence of *B. anthracis* Ames (FBI/USDOJ, 2011, FBI Documents, WFO Report). This work was performed as part of the anthrax letters investigation. Few details were made available to the committee.

At least three sample collection missions were conducted by the FBI and/or partners from the intelligence community at an overseas site because of information about efforts by Al Qaeda to develop an "anthrax program" (FBI/USDOJ, 2011). In May 2004, the FBI and partners from the intelligence community visited an overseas location at which they had been told an anthrax program had been operating, and brought back swab and swipe samples to the United

States. None of the samples grew *B. anthracis* after incubation in culture media. However, three swab samples were reported as positive for *B. anthracis* and for *B. anthracis* Ames-specific sequences by PCR, including swabbings from the outside of an unopened medicine dropper package, a sink, and a sink drain hose. Repeat testing of these three positive samples as part of a group of 15 blinded samples, including soil samples, water blanks and non-Ames *Bacillus* species, again yielded positive results for two of the three same samples (and for none of the other samples). However, not all replicates of the DNA extracts from the positive samples gave positive results. Apparently, an earlier collection mission to this site, prior to May 2004, by others in the intelligence community had also yielded samples with positive PCR results for *B. anthracis* DNA and negative culture results. As a result of these findings, a third collection mission was conducted in November 2004 and this time large portions of the site were returned intact to the United States, including the entire sink, drain, and associated plumbing that had been the source of the positive March 2004 samples. These items were extensively sampled, and again tested for both viable *B. anthracis* and for *B. anthracis* DNA. This time, according to the June 2008 declassified document, all the tests were negative (FBI Documents, WFO Report).

The committee was provided only fragmentary information about and limited primary data from this work and received them very late in our study. We consider these data to be inconclusive regarding the possible presence of *B. anthracis* Ames at this undisclosed overseas site. Several scientific and technical issues should be explored in more detail, such as the performance characteristics of the assays, whether or not the assays were validated for use with these sample types, the degree to which samples or sample locations gave repeatedly positive results, interpretation of inconsistent positive results, whether or not the Ames genetic mutations in the anthrax letters were detected in any of these overseas samples, and the natural distribution of *B. anthracis* strain types in this overseas geographic region.

3.4.4 Letter Material and Cross Contamination

Material was collected directly from the Daschle, Leahy, *New York Post*, and Brokaw letters, but in varying quantities. As indicated above, no letter was found at the AMI building in Florida, thus only environmental and clinical samples were available for analysis. The decision by the U.S. Attorney's Office not to pursue molecular analysis on the AMI crime scene samples (FBI, 2011) limits the ability to definitively connect this attack to the material in the recovered letters from New York and Washington, D.C.

The letter addressed to Tom Brokaw at NBC was found after Erin O'Connor, an assistant to Brokaw, developed cutaneous anthrax after opening a letter containing a white powder (Cole, 2009). As mentioned before, the FBI laboratory was not equipped to handle *B. anthracis*, so all of the New York samples were sent to the New York Department of Health in Albany or elsewhere. The New York City Health Department officials who tested the Brokaw letter accidentally lost most of the sample and contaminated the laboratory, rendering the space unusable for a critical period of time after the *B. anthracis* was discovered and before the full scope of the problem was known. As a result, little material from the Brokaw letter remained available for further analyses. A biopsy obtained from a black eschar lesion that formed on O'Connor and was sent to CDC tested positive by immunohistochemical staining for the cell wall antigen of *B. anthracis*.

The *New York Post* letter had been thrown into a bin for hate mail and was found after employee Johanna Huden became ill with cutaneous anthrax. The letter had been unopened and the enclosed material was available for future analyses.

On October 15, when staff in Senator Daschle's office opened an envelope containing a white powder, police quarantined the office and surrounding rooms, shut down the Capitol's mail system, and suspended public tours. Law enforcement officials in protective biohazard suits took over Senator Daschle's office. Unlike previous samples, which were sent to the CDC for analysis, the FBI sent the Daschle letter to USAMRIID because of its biocontainment facilities (FBI, 2009). Since much of the material was dispersed when the letter was opened, limited material was available for future analyses.

On November 16, 2001, FBI and EPA HAZMAT personnel found the letter addressed to Senator Leahy in one of 280 barrels of unopened mail collected from Capitol Hill after the discovery of the anthrax-contaminated letter sent to Senator Daschle (FBI, 2008a). The search that resulted in retrieval of the Leahy letter was conducted by teams of HAZMAT workers from the FBI and EPA Criminal Investigative Division (FBI, 2008a). They developed a sampling protocol intended to eliminate the need for HAZMAT teams to sift through each piece of mail in 642 trash bags to find contaminated mail (FBI, 2008a).

The mail was sampled and sorted in a containment facility constructed in a large warehouse and maintained with negative air pressure. The intake and exhaust air was passed through HEPA filters, which trap essentially all particles the size of anthrax spores. Air samplers were used to monitor the air inside and outside the containment area for the presence and quantity of airborne spores. Investigative workers were monitored for exposure by sampling their clothing. In general, the only investigators who were contaminated were those who handled a "hot" bag containing the letter addressed to Leahy laden with spores (Beecher, 2006).

Each bag was shaken in an attempt to distribute any spores that could be present. A swab was then inserted into a small hole in each bag and wiped around the inside. After the swab was withdrawn, the hole was sealed with duct tape and the swab was used to inoculate a Petri dish, which was sent to the Naval Medical Research Center (NMRC) for analysis (FBI, 2010b).

Trace contamination was detected in 62 bags, likely due to a high level of shedding from cross-contaminated mail (Beecher, 2006). Five bags produced more than 100 bacterial colonies from a swab and were considered "hot." Innovative air sampling was used to maximize the recovery of the target organism (Beecher, 2010). One bag, the one containing the Leahy letter, was orders of magnitude more contaminated than the others. This bag produced between 19,000 and 23,000 spores (or 760 to 920 colony-forming units per liter of air sampled) (Beecher, 2006).

Beecher (2006) examined some of the issues related to cross contamination of mail, mail bags, and the local environment by the Leahy letter. He documented extensive contamination of personnel especially during physical handling of the spore-laden letter. In the conduct of this work, the investigators followed thoughtful and appropriate practices and procedures for the purpose of minimizing artifactual cross contamination. Of note, they identified a correlation between the degree of letter contamination and the previous handling of the letter by sorting machines in the Trenton and Washington, D.C., postal processing and distribution centers. The insights from this work proved useful in developing a more detailed understanding of the route taken by the anthrax letters from the point of deposit through the U.S. Postal System to the site of delivery. Although suggestive of a mechanism and scenario by which Ottilie Lundgren, a resident of Oxford, Connecticut, might have developed inhalational anthrax (cross contamination of mail en route to her home), the committee lacked sufficient information (e.g., other possible exposures, unusual susceptibility to low numbers of anthrax spores) with which to assess the plausibility and likelihood of this mechanism and this particular scenario.

On September 1, 2001, Defence Research Establishment Suffield (DRES) in Canada released the results of a study (FBI Documents, B2M11D1) that had been designed to measure and better understand the dispersion of spores that might occur after the opening of an envelope containing *B. anthracis* spores. This study involved a series of experiments in which envelopes containing either 0.1 or 1.0 gram of *B. globigii* spores at a concentration of $\sim 1 \times 10^{11}$ cfu/g (as a surrogate for *B. anthracis* spores) were opened in a DRES aerosol test chamber that was configured to represent a mail room or office. The chamber measured $18 \times 10 \times 10$ ft (i.e., with a volume of 1,800 cu ft) and had a recirculating air handling system operating at 1,050 cu ft/min. The presence of spores at various sites in the chamber was assessed using culture-based approaches, not molecular detection methods. The results showed that the act of handling or opening these envelopes was "far more effective than initially suspected" in causing dispersion of spores in the chamber. Particles of respirable size were released quickly and spread throughout the chamber, such that after the opening of a 0.1 g spore envelope, 10 minutes of exposure to the air in the chamber would have provided a dose 480-fold greater than the amount needed to kill a human with 50 percent probability. The investigators noted that envelopes were more likely to cause cross contamination of the local environment, including the envelope handler, if the open corners of the envelope were not deliberately sealed by the preparer. The investigators at DRES did not seal the corners of the envelopes they used in these experiments. However, the corners of the envelopes mailed through the U.S. Postal System with *B. anthracis* spores in September and October 2001 apparently were sealed. This study was valuable in revealing the potential speed, magnitude, and spatial distribution of environmental contamination by spores subsequent to the handling or opening of a spore-laden envelope.

3.5 COMMITTEE FINDINGS AND RECOMMENDATIONS

The events of autumn 2001 unfolded rapidly, with CDC's initial public health response quickly extended to a major criminal investigation under the control of the FBI. A multipronged investigative strategy emerged for the scientific investigation, with one set of activities focused on understanding the characteristics of the material in the letters and another on developing and conducting a comparative analysis of these evidentiary materials against samples collected from the scientific community. The science and technology that formed the basis for these analyses evolved rapidly and had a major impact on the field of microbial forensics. Although the public health crisis largely subsided after the last victim died in November 2001, the scientific investigation continued until 2008 and the criminal investigation continued until the case was closed in February 2010 (see Table 3-1).

Finding 3.1: Over the course of the investigation, the FBI found and engaged highly qualified experts in some areas. It benefited from the unprecedented guidance of a high-level group of agency directors and leading scientists. The members of this group had top secret national security clearances, met regularly over several years in a secure facility, and dealt with classified materials. The NRC committee authoring this report, in keeping with a commitment to make this report available to the public, did not see these materials.

In a complex investigation during a period of extreme national urgency such as this one, it is imperative to recruit and make wise use of the best and most relevant expertise. The Bureau regularly briefed its high-level advisory group in a secure setting where classified material was reviewed. They also relied on subject matter experts and expertise from other government agencies. The committee recognizes, in agreement with the FBI (FBI/USDOJ, Jan 2011) that the unique skills and expertise needed to conduct microbial forensic examinations might, in some cases such as this one, require reliance on the same set of scientific personnel, who also might be considered potential suspects. Thus it is important to have in place an external oversight structure with the capacity to recognize where relevant expertise and conflicts might reside and to provide guidance and coordination for the overall scientific investigation. In addition, the small number of laboratories with expertise in a given pathogen or technique required that a certain level of independent oversight be maintained.

Finding 3.2: A clear organizational structure and process to oversee the entire scientific investigation was not in place in 2001. In 2003, the FBI created a new organizational unit (the Chemical, Biological, Radiological, and Nuclear [CBRN] Sciences Unit, sometimes referred to as the Chemical Biological Science Unit, or CBSU) devoted to the investigation of chemical, biological, radiological, and nuclear attacks. The formation of this new unit with clearer lines of authority is commendable.

Weekly meetings occurred between the science team and those leading the criminal investigation. The results of these meetings and the way the scientific, criminal and legal aspects of the investigation interplayed will be beneficial in preparing for future attacks (B3D1). In addition, the development of a new U.S. Government microbial forensics infrastructure and research strategy will be important steps toward enhancing future capabilities for attribution in the event of a biological attack (Pesenti, 2010).

Finding 3.3: Investigators used reasonable approaches in the early phase of the investigation to collect clinical and environmental samples and to apply traditional microbiological methods to their analyses. Yet during subsequent years, the investigators did not fully exploit molecular methods to identify and characterize *B. anthracis* directly in crime scene environmental samples (without cultivation). Molecular methods offer greater sensitivity and breadth of microbial detection and more precise identification of microbial species and strains than do culture-based methods.

The committee recognizes that the circumstances of 2001 created an abundance of samples and associated work for the LRN (laboratory response network) resulting in a decision to use RODAC plates for analysis of the environmental samples. Thus, the FBI did not confirm the presence of the letters-associated genetic mutations in the environmental samples. In the nine years since that time, dramatic advances in high throughput sequencing technology have greatly improved the ability to detect and characterize rare strains and species in complex environmental samples[2] (for further, related discussion,

---

[2] Multiple studies in each of a number of diverse environmental settings have demonstrated the feasibility and reliability of resolving strain-specific fine genomic structure (at the level of SNPs and local structural rearrangements) in highly complex biological samples using next-generation sequencing technology and metagenomic approaches. Examples include strain resolution and gene

see Finding 6.8 in Chapter 6). In thinking of future incidents, it is imperative, based on lessons learned from this investigation, to anticipate the types of situations and circumstances that might affect evidence collection, preservation, and documentation, and to employ protocols and procedures that ensure the best possible outcomes.

Finding 3.4: There was inconsistent evidence of *B. anthracis* Ames DNA in environmental samples that were collected from an overseas site.

At the end of this study, the NRC committee was provided limited information for the first time about the analysis of environmental samples for *B. anthracis* Ames from an undisclosed overseas site at which a terrorist group's anthrax program was allegedly located. This site was investigated by the FBI and other federal partners as part of the anthrax letters investigation. The information indicates that there was inconsistent evidence of Ames strain DNA in some of these samples, but no culturable *B. anthracis*. The committee believes that the complete set of data and conclusions concerning these samples, including all relevant classified documents, deserves a more thorough scientific review.

Finding 3.5: As was done in the anthrax investigation, at the outset of any future investigation the responsible agencies will be aided by a scientific plan and decision tree that takes into account the breadth of available physical and chemical analytical methods. The plan will also need to allow for possible modification of existing methods and for the development and validation of new methods (see Chapter 4, Section 12).

The scientific investigation of any future biological attack would greatly benefit from robust independent oversight and ongoing review. To accomplish this, the government should maintain a standing body of scientific experts with proper security clearances who are fully briefed on matters of importance for preparedness and response to a biological attack. When an investigation is launched, members of this group could help guide the scientific investigation.

In preparing for future investigations, all relevant U.S. government agencies and departments will need to work together to ensure that independent,

---

family structure in ocean surface water (see multiple publications from the *Sorcerer II* Global Ocean Sampling Expedition), evidence for rare strain ecotypes in acid mine drainage (see multiple publications from the Banfield group at UC Berkeley), and gene family sequence microheterogeneity in prophage genomes that have been reconstructed from shotgun sequencing of virus-like particles in human feces (e.g., see supplemental data from *Nature* 466:334, 2010). This large and growing body of work strongly suggests that the application of these same techniques to some of the environmental samples from the anthrax letters case might provide additional clarity about *B. anthracis* genome sequence variants and the relationships among the strains in the samples and the letters.

high-quality, external science advice is available from individuals with expertise in critical scientific areas likely to be relevant given anticipated scenarios for scientific investigations. By identifying a core set of external experts and convening this group ahead of time, productive working relationships could be established between this group and members of the government's bioforensics community. When a new investigation is launched, this core set of external experts could assist in recruiting others with more specific expertise relevant to the investigation under way. For example, the FBI included some expert scientists early in the anthrax letters investigation, but it does not seem to have sought formal expertise in statistics until the investigation was nearly completed. Because many inferences depend on the design and analysis of datasets that may be complex, as in this case, for any similar investigation in the future, it will be important that the FBI consult with expert statisticians throughout the processes of experimental design and planning, sample collection, sample analysis, and data interpretation. The committee recognizes that much has been done by the government over the past several years to build and enhance this important infrastructure and set of resources, although the scope of this study did not include assessment of the current infrastructure.

When an investigation is launched, the panel of external experts could recommend and review strategies, protocols, and procedures; help with the development of new methods and scientific approaches; provide advice on the selection of contract scientists and additional outside experts; assist in data interpretation; and help generate alternative hypotheses. Members of the panel should not be directly involved in conducting the scientific investigation itself. It will also be important that the panel have an ongoing role throughout the course of the investigation and be briefed on all of the science that is contemplated and pursued. Consistent documentation of recommendations and input, and the subsequent responses, will be of great value.

In the future relevant agencies should review and periodically update appropriate protocols and experimental designs to use best strategies for preserving evidence, exploiting samples for scientific information, and meeting subsequent legal challenges. Protocols should ensure that clinical and environmental evidence is properly collected, preserved, documented, and analyzed to maximize the utility of the samples collected. Furthermore, state-of-the-art molecular methods, such as next-generation nucleic acid sequencing techniques, offer great potential for characterizing clinical and environmental samples (see Finding 6.8 in Chapter 6).

Recommendation 3.1: A review should be conducted of the classified materials that are relevant to the FBI's investigation of the 2001 *Bacillus anthracis* **mailings, including all of the data and material pertaining to the overseas environmental sample collections.**

The committee did not receive nor review classified material. In November 2010 discussions with FBI and DOJ leadership regarding this report, we were made aware of additional information that would require review of classified material. Due to the lateness of this revelation and the importance we placed on issuing a timely report, and the agreement between the NRC and the FBI that all material we considered be publicly available, the committee did not undertake this additional review of classified material.

Recommendation 3.2: The goals of forensic science and realistic expectations and limitations regarding its use in the investigation of a biological attack must be communicated to the public and policymakers with as much clarity and detail as possible before, during, and after the investigation.

Communicating with the public and policymakers is extremely important in order to ensure that accurate information is available and to minimize unrealistic expectations. Special attention will need to be paid to communicating scientific information to these groups in an accurate and credible manner, especially if the information will play a critical role in the investigation.

When presenting to the public the findings of an investigation that involve scientific evidence, especially one as important as the anthrax letters investigation, officials will need to make every effort to have scientists verify the accuracy of the scientific information they report. The inaccurate reporting of facts or the overstatement of scientific evidence is a disservice to the public. In the anthrax letters investigation, there were repeated claims that all of the attack letters contained all of the genotypic variants (see Chapter 6 and Finding 6.7) that implicated flask RMR-1029 as the source of the anthrax spores, when in fact not all of the letters were checked for these variants. Of even greater concern, because it suggested possible deception by the suspect, the strength of the evidence was overstated that a disputed sample submitted by the suspect had not come from the proper source (see Chapter 6 and Finding 6.4). Similar mistakes can be avoided in the future by involving the relevant scientists in fact checking of the reports before they are released.

4

Physical and Chemical Analyses

4.1 INTRODUCTION

Early in its investigation of the 2001 *Bacillus anthracis* (*B. anthracis*) attacks, the FBI hired several laboratories and conducted some of its own analyses to ascertain physical and chemical characteristics of evidentiary materials, as summarized in Table 3-1 (Chapter 3). These analyses of the letter powders focused on the size and granularity of the particulates, elemental content, age of the spores, and identification of chemical signatures that might provide clues related to the source or production processes used.

The physical and chemical analytical methods used by the FBI and outside contractors were conducted properly and most were well established at the time and validated for use in law enforcement investigations. In some instances, new methods were developed to accomplish the desired analytical measurements. Other methods, such as mass spectrometric analysis and various microscopy techniques, are well accepted but were applied in new ways and for new target analytes in this investigation (e.g., heme, agar, and additive analysis, isotopic analysis of letter evidence, silicon determination in spores). This chapter describes and evaluates the chemical and physical analyses performed.

4.2 SPORE PREPARATION AND PURIFICATION

One line of inquiry in the FBI investigation concerned the expertise, time, and technology needed to produce the material used in the attacks. That is, what skills, tools, or procedures would be needed to cultivate, purify, and dry the spores, and how long would the entire process take? While aspects of this issue pertain to the traditional criminal component of the investigation, other aspects pertain to the science component. The committee restricted its review and analysis to these latter aspects. Committee efforts were complicated by the fact that the FBI did not publish or provide the committee with specific or detailed conclusions on its theories regarding the methods used for cultivating, purifying, or drying the spores found in the letters. In discussions with the FBI at the January 2011 meeting, the FBI stated that some of its consulting experts referred to the letter preparations as being of "vaccine quality", which narrowed the list of potential suspects. Nonetheless, the Bureau investigated individuals without regard to their specific skill sets. The FBI further stated that the time for preparation and equipment used in preparation of the letter materials was difficult to ascertain because of numerous variables. However, FBI officials indicated that inferences about required skills or time for spore preparation were never the sole criterion for eliminating suspects (FBI/USDOJ, 2011).

With regard to cultivation of the spores in liquid versus solid (i.e., agar-based) medium, assays for the presence of residual agar in the letter material were inconclusive, as described below. The DOJ *Amerithrax Investigative Summary* (USDOJ, 2010) describes the potential use of either a fermentor or an incubator with shaken flasks and liquid media. That document also suggests that a minimum of 500 ml of liquid culture would be required to produce the spores in the letters but then states "we cannot say with certainty how much material was used in the letters."

Given the information available on the number of spores believed to have been placed in the letters and knowledge of spore yield from various types of cultivation methods, a range of required culture volumes can be estimated. Four spore-containing letters were recovered and evidence indicates that one additional letter was sent to American Media, Inc. (AMI) (Cole, 2009). A high estimate for the total number of spores sent through the mail would include five letters, each containing 1 gram of spore-containing powder with 2×10^{12} spores per gram, for a total of 1.0×10^{13} spores. A low estimate for the total number of spores sent through the mail would include five letters with 0.8 gram of spore-containing powder per letter. Two of the letters (Leahy and Daschle) might have contained 2×10^{12} spores per gram while the others (*New York Post*, Brokaw, and AMI) might have contained 2×10^{11} spores per gram (FBI Documents B1M2D1, B1M2D3, B1M2D6), for a total of 3.7×10^{12} spores (see Table 4-1).

TABLE 4-1 Estimated Ranges of Total Number of Spores

Total spores contained in all letters	Number of letters, gram spores per letter, and spores/gram
Low estimate = 3.7×10^{12}	2 letters with 0.8 gram spores per letter and 2×10^{12} spores per gram
	3 letters with 0.8 gram spores per letter and 2×10^{11} spores per gram
High estimate = 1.0×10^{13}	5 letters with 1 gram spores per letter and 2×10^{12} spores per gram

A published yield of *B. anthracis* Ames spores grown on solid medium is 8×10^9 spores per Petri 150-mm plate (Carrera et al., 2007). Production of the required number of spores on plates, with the conservative assumption of no spore losses during purification, would therefore require 463 to 1,250 plates. Expert testimony to this committee indicated that 15 liters of liquid culture in a fermentor could yield 2×10^{13} *B. anthracis* spores (Heine, 2010). Cultivation in shake flasks or losses during spore purification could certainly reduce this yield severalfold. RMR-1029, was a well-characterized large-scale spore preparation housed at the U.S. Army Medical Research Institute for Infectious Diseases (USAMRIID). The initial 1 liter purified spore preparation in RMR-1029 was derived from approximately 160 liters of culture and contained an estimated 3×10^{13} spores. Thus, cultivation in the range of 2.8 to 53 liters of liquid medium would have been required to produce the spores required for the letters (see Table 4-2).

Spore purification is typically accomplished by repeated centrifugation, disposal of the supernatant-containing cellular debris, and resuspension of the spore pellet in fresh liquid. High spore purity is generally achieved by centrifugation of the spores through a gradient of a high-density compound. The most commonly used high-density compound, meglumine diatrizoate, was not detected in the letter material (see Section 4.6 below), suggesting that this procedure was not used or that the spores were extensively washed after such a procedure. Purification by any method would involve some liquid washing steps and would require a relatively large-capacity centrifuge. Such instruments are commonly found in microbiology laboratories. The spores in the Leahy and Daschle letters were accompanied by less contaminating debris, and thus were of higher purity and concentration, than those in the *New York Post* letter (FBI Documents, B1M2D2, B1M2D6).

There are several methods for drying spore suspensions to produce powders like those found in the letters: these include chemical desiccation, air drying, and freeze drying (lyophilization), any of which could require several hours to several days. Drying of surrogate spore preparations using various

TABLE 4-2 Estimates of Media Volume Required for Spore Preparation

Total spores needed for all letters	Spores prepared on plates at 8×10^9 spores/plate	Spores prepared in liquid
Low estimate = 3.7×10^{12}	463 plates	2.8 liters in a fermentor, based on Heine's (2010) estimate of 2×10^{13} spores from 15 liters
High estimate = 1.0×10^{13}	1250 plates	53 liters in a fermentor, based on RMR-1029 having 3×10^{13} spores from 160 liters

methods produced particle size distributions similar to those found in letter samples (described below), but volatile organic compounds that might have been used for drying were not detected at significant levels in letter samples (see Section 4.8 below). The FBI therefore offered no conclusions concerning the method used to dry the attack spores.

As a result of the different possible production schemes that might have yielded product with the observed characteristics of the evidentiary materials, the committee finds that the time required for this work could be as little as 2 or 3 days to as much as several months. The differences are based on different estimates of the time required for propagation, purification, and drying, among other variables, as well as the state of the starting material. In particular, it is not known whether some of the initial steps might have occurred well in advance of the letter attacks. The committee cannot resolve these distinctions because it had no information identifying a production method or the steps involved in production.

The FBI did not present a definite theory on how and when propagation, purification, and drying took place, nor on what specific skills would be required to perform these tasks. Nonetheless, inferences made by the FBI concerning the time, skill, and equipment required for spore preparation were said to be significant considerations in its narrowing of the list of potential suspects (USDOJ, 2010, pp. 29-33, 36-38), but were never the sole criteria for eliminating suspects (FBI/USDOJ, 2011). Without further specification with respect to spore preparations variables, the committee finds no scientific basis on which to accurately estimate the amount of the time or specific skill set needed to prepare the spore material.

4.3 SURROGATE PREPARATION AND PURIFICATION

In pursuit of greater clarity on the issues discussed above, the FBI asked scientists at Dugway Proving Grounds (DPG) to create surrogate samples in an attempt to mimic the physical properties of the letter samples. Chemical properties of the surrogates were also compared to those of the letter samples. Surrogates designated in the Weekly Updates as "Buran" and "Abshire" were characterized with respect to physical and chemical properties, but the preparation procedures for these samples were not provided to the committee. The committee did not learn of the scientific rationale behind the choice of the various preparations studied, other than an apparent effort to study a wide range of preparation conditions.

Surrogates were grown by both plate and fermentation methods (using *B. anthracis* from the Leahy letter as the starter source) (FBI Documents, B1M13D3). Plate methods included various combinations of growth media (sheep blood agar, new sporulation medium), washing (DPG, Patrick methods), drying (oven/air, lyophilizer, acetone, speed vac), and milling (ball mill, mortar and pestle, sieve)—36 preparations in all. Fermentation methods included two types of antifoam agents, one containing silicon in the form of polydimethylsiloxane (Antifoam C) and one silicon free (Antifoam 204) (DPG, 2006). A separate set of *B. anthracis* Sterne preparations was analyzed at Lawrence Livermore National Laboratory (LLNL) beginning in 2006 for the specific purpose of studying silicon uptake into the spore coat. In situations where it was inconvenient or unsafe to work directly with *B. anthracis*, surrogate samples of other *Bacillus* species were prepared. The results of these analyses are discussed below.

4.4 SIZE AND GRANULARITY OF THE MATERIAL IN THE LETTERS

In fall 2001 at the request of USAMRIID, the Armed Forces Institute of Pathology (AFIP) performed scanning electron microscopy (SEM) to determine the size and shape of particulates in the letter material. SEM is a common microscopic imaging technique used extensively across scientific disciplines. Results of this analysis demonstrated that the size and shape of spores found in the letters were consistent with *B. anthracis* (FBI Documents, B1M2D4; AFIP, 2001). The images showed individual spores as well as clusters of spores, and other solid (crystalline) material such as calcium carbonate (Kuhlman, 2001b). While the morphology may have been affected by sample preparation prior to analysis (autoclave), the images were consistent with measurements of particle size.

In fall 2001, Battelle Memorial Institute (BMI) evaluated size distributions of aerosolized particles from untreated letter material and surrogate samples to determine whether respirable-size particles were present and whether the amount of such particles would have required specialized protocols for preparation (e.g., dispersants).[1] Well-established commercial instruments were used to aerosolize the samples (AeroDisperser®, TSI, Inc.) and measure their particle size distributions (Aerosizer®, TSI, Inc.). The AeroDisperser uses a high-velocity gas to lift and disperse solid particles (powders) off a plate into the airflow. The Aerosizer measures the aerodynamic diameters of individual particles by accelerating the aerosol in a sonic airflow and determining the velocity by time of flight between two laser beams. The Aerosizer was calibrated with National Institute of Standards and Technology traceable particle size standards between 5 and 20 micrometers (μm) in diameter.

The Daschle and Leahy letter samples had bimodal particle size distributions, with one mode around 1.5 μm in diameter, corresponding to the size of an individual *B. anthracis* spore, and another mode greater than 20 μm in diameter, corresponding to the size of clusters of large numbers of spores and other material

---

[1] To remove the implication that autoclaving of letter samples was standard practice, the word "untreated" has been inserted into this sentence. The insertion represents a modification to the text that appeared in the prepublication edition of this report.

(FBI Documents, B2M13D11). In the Daschle sample, 0.05 percent of the total volume (mass) of particles was found in the smaller diameter mode. In the Leahy sample, 1 percent of the total mass was found in the smaller diameter mode.

Several *Bacillus subtilis* var. niger culture preparations made using only centrifugation for concentration and lyophilization for drying also had bimodal size distributions, with the smaller (1.5 µm diameter) mode constituting approximately 1 percent of the total aerosolized mass (Kuhlman, 2001a,c). The similarity between the letter and these size distributions showed that powders with dispersion characteristics similar to those of the letter material could be made without the addition of a dispersant.

Particle size distributions for the Dugway surrogate samples were reported as mean particle diameter, which unfortunately is a less informative indicator of particle size when the distribution is bimodal. Nonetheless, many of the Dugway preparations gave mean particle diameters in the same range as the letter samples, 2 to 4 µm, consistent with the notion that dispersants were not required to produce powders with these particle size distributions (DPG, 2006).

A recent report in the scientific literature describes production of *B. anthracis* spores in a manner to enhance formation of particles of about 1.5 µm in diameter (Baron et al., 2008). The Baron preparation involved a proprietary drying procedure, ball milling, and the addition of 20 percent amorphous silica fluidizing agent. The size distribution showed a spore mode around 1.5 µm in diameter and a smaller mode around 0.5 µm in diameter. It is not possible to compare the particle size distributions of the letter samples with the Baron work since the latter study included an impactor to remove particles larger than about 5 µm.

4.5 PRESENCE OF SILICON AND OTHER ELEMENTS IN THE LETTER MATERIAL

While any deliberate mailing of letters containing anthrax spores might be considered a form of spore weaponization, this term has been more commonly used to describe preparations with enhanced properties of dispersion and aerosolization. It is commonly believed that deliberate efforts to make a powder more dispersible through the use of additives would suggest a more sophisticated level of preparation expertise. Thus the presence of dispersants, such as nanoparticulate silica or bentonite, was an important feature in considering whether or not the letters contained "weaponized" anthrax spores. The FBI commissioned several studies to determine whether silicon was present in the letters and, if so, to ascertain the nature of the silicon.[2]

---

[2] The elemental analysis methods discussed in this chapter determine the amount of silicon in the sample but provide no direct information about the chemical form of this element other than what can be inferred from simultaneous detection of other elements. Therefore, the term "silicon" is used in reference to the results of elemental analysis.

4.5.1 Elemental Analysis

Elemental analysis can be performed in several ways, providing varying levels of sensitivity and spatial resolution. The FBI worked with laboratories having several capabilities and these techniques are summarized in Table 4-3. One approach is to measure the presence of heavy elements while performing electron microscopy. As stated previously, AFIP performed SEM in fall 2001 to characterize the morphology of the letter samples. A complementary tool in many scanning electron microscopes, energy-dispersive X-ray analysis (SEM-EDX), provides a low-resolution analysis of the spatial distribution of elements in the samples. SEM-EDX is a well-established analytical technique that monitors the X-ray emission from the small region of sample that is irradiated by electrons. The wavelengths of X-ray emission identify the elements present, while the corresponding intensities permit quantification of each element in the irradiated region. These initial measurements showed high levels of silicon in the letter material (USAMRIID, 2001).

In early 2002 the FBI performed elemental analysis on letter and surrogate samples using inductively coupled plasma-optical emission spectroscopy (ICP-OES) (FBI, 2009). ICP-OES is a well-established chemical analysis tech-

TABLE 4-3 Methods for Chemical Analysis Referred to in Chapter 4

Method	Primary use in the investigation
Inductively coupled plasma-optical emission spectrometry (ICP-OES)	Bulk elemental composition of a sample dissolved in aqueous solution
Scanning electron microscopy with energy dispersive X-ray analysis (SEM-EDX)	Size-resolved elemental composition of a solid sample, with size resolution of about 1 µm or less
Nanometer secondary ion mass spectrometry (nano-SIMS)	Size resolved elemental and molecular composition of a solid sample, with size resolution of about 1 µm or less
Accelerator mass spectrometry (AMS)	Measurement of the ratio of ^{14}C to ^{12}C in a sample
Isotope ratio mass spectrometry (IRMS)	Measurement of the ratios of ^{2}H to ^{1}H and ^{18}O to ^{16}O in a sample
Gas chromatography-mass spectrometry (GC-MS)	Separation and identification of volatile organic compounds (i.e., compounds that can be vaporized without decomposition)
Liquid chromatography-mass spectrometry (LC-MS)	Separation and identification of polar, nonvolatile organic compounds
Aerodynamic particle sizer (Aerosizer®)	Size distribution of an aerosol, by time of flight in sonic flow
Aerodisperser	Size distribution of an aerosol particle by lift

nique that provides accurate and precise measures of bulk elemental composition (weight percentages of individual elements in the sample), but gives no information about the spatial distribution of the elements in the sample. ICP-OES analysis showed quite high weight percentages of silicon (Table 4-4): 10 percent in the *New York Post* material and 1.4 percent and 1.8 percent for two separate samples of the Leahy material. Control samples of *B. subtilis* with and without the addition of a dispersant, were also analyzed, and the sample containing dispersant showed a higher measured silicon content than matched samples without dispersant.

The large amount of silicon in the letter materials fueled speculation that dispersants such as silicon dioxide or bentonite (an aluminum silicate clay) may have been present, possibly indicating intent to "weaponize" the material by enhancing its ability to disperse. A control sample of bentonite was analyzed as part of the original AFIP work for USAMRIID (FBI Documents, B1M2D4). Bentonite was ruled out rather quickly as an additive to the material from the *New York Post*, Leahy, and Daschle letters owing to the lack of a commensurate amount of aluminum and other elements in the SEM-EDX images. Silicon dioxide remained a possibility and was investigated using SEM-EDX to probe the spatial distribution of elements in the samples.

TABLE 4-4 Summary of Silicon Measurements in Evidentiary and Surrogate Samples

Sample[a]	Bulk silicon content ICP-OES	SEM-EDX Analysis	SEM-EDX % spores with Si in coat
Leahy	1.4-1.8%	1-2% per spore	97/124 = 76%
Daschle		1-2% per spore	73/111 = 66%
New York Post	10%	1-2% per spore	91/141 = 65%
RMR-1029			0/98 = 0% 0/115 = 0% 0/191 = 0%
RMR-1030 (shake flask preparation)			6/94 = 6% 0/118 = 0% 7/113 = 6%
Dugway surrogates	0.2-5% (10 agar preparations)		42/163 = 26% 17/161 = 11% 50/172 = 29% (fermentation using Leighton-Doi media) 0% (2 agar preparations)

NOTES: ICP-OES = inductively coupled plasma-optical emission spectrometry; SEM-EDX = energy dispersive X-ray analysis; SI = silicon.

[a]Similar analysis of Brokaw letter material was not performed due to the small amount of sample available.

Bulk elemental analysis by ICP-OES of surrogate preparations from Dugway, without the addition of a dispersant, was reported for 10 of the 36 preparations (FBI Documents, B1M7). All of the surrogates analyzed were found to contain silicon at a level of 0.2 weight percent or more, and four contained silicon in the 2- to 5-percent range. These results indicate that it is possible to prepare spores having a silicon content in the range of the Leahy letter sample without adding a dispersant (FBI Documents, B1M7).

Trace levels of several elements including aluminum, sodium, magnesium, potassium, calcium, chlorine, iron, manganese, zinc, and tin were detected in letter and surrogate samples as well as in preparation reagents. While some of these elements are ubiquitous impurities in the environment, others are less common and sometimes are used in chemical profiling in forensic investigations. The Amerithrax Science Update documents (B3D1) describe multiple efforts to link stainless steel fragments contained in letter material with production methods or other sources of stainless steel particles (e.g., drugs, growth media components, pens). As of 2005 Michael and Kotula of Sandia National Laboratories believed that the tin and iron present in the powders may have provided a useful chemical signature; however, the committee was never shown any evidence to indicate that this possibility was pursued further (FBI Documents, B1M1D7) or that these discussions led to any conclusions about the source of material or production methods.

4.5.2 Spatially Resolved Elemental Analysis

Sandia National Laboratories (SNL) offered instrumentation with better spatial resolution and detection sensitivity than the instruments at AFIP and FBI. Beginning in February 2002, the FBI sought SEM-EDX analyses of letter and surrogate samples using instruments and algorithms developed by staff scientists at SNL that elucidate correlations among the spatial distributions of different elements. These enhanced capabilities greatly aided the measurements and the conclusions that could be drawn from them.

SEM-EDX can be performed in low- or high-spatial resolution mode depending on how tightly the electron beam is focused. Low-resolution operation can provide an estimate of the bulk elemental composition of a sample, although it is a less reliable indicator of bulk composition than ICP-OES. Low-resolution SEM-EDX analysis of both the *New York Post* and Leahy samples gave estimates of 1 to 2 percent for the bulk silicon content. The consistency between ICP-OES and SEM-EDX measurements (Table 4-4) of the Leahy material, which appeared to be a refined and homogeneous powder, argues against the existence of a method-dependent artifact that would lead to an incorrect measurement by one method or the other. On the other hand, the ICP-OES and SEM-EDX measurements of the *New York Post* material differed by an order of magnitude. Further complicating the comparison, the

2001 AFIP SEM-EDX images of the *New York Post* material (obtained by the FBI during the course of this study) showed regions enriched with silicon but not oxygen, suggesting the presence of a reduced form of silicon (FBI Documents, AFIP, 2001). These regions were apparently missing or not reported in the SNL SEM-EDX images. When asked at the January 2011 meeting, the FBI stated that the presence of reduced silicon was "just an observation" and that differences between the two methods (AFIP and SNL) could account for the discrepancy as one of the methods relied on sectioned spores while the other did not.

4.5.3 Silicon in the Spore Coat

High-resolution SEM-EDX of the Leahy and *New York Post* samples gave similar spatial distributions—silicon was concentrated within the spore coat, but none was detected in the exosporium where a dispersant would reside (Figure 4-1). While the SEM-EDX images appear convincing, two questions should be considered: (1) Is silicon in the exosporium detectable in the experiment, or might it be lost during sample preparation? (2) How was silicon incorporated into the spore coat and does its presence in the samples have forensic value?

The FBI approached the first question by preparing a sample of *B. subtilis* spiked with 20 percent Syloid 244, a commercial nanoparticulate silica product, to simulate a flow-enhanced sample. When this sample was mounted and analyzed in the same manner as the letter samples, silicon dioxide nanoparticles were clearly observed on the spore surface. Although this experiment indicates that a dispersant would have been detected if it were present in the sample, a more definitive experiment would have been to spike an actual letter sample with dispersant at levels comparable to the bulk silicon content. The latter experiment was not performed.

Initially, the FBI provided evidentiary samples to SNL that were already fixed and stained. It is unclear whether or not this preparation would have resulted in separation of the dispersant from the spores prior to analysis. The FBI subsequently provided letter samples to SNL that had been treated only by gamma irradiation, which should not have resulted in loss of dispersant. The irradiated samples also showed silicon only in the spore coat.

Insight into the second question can be gained from a study showing that *B. cereus* spores accumulate silicon in the coat (Stewart et al., 1980). The FBI obtained archived samples from that study and had SNL analyze them by SEM-EDX. The images showed silicon in the spore coat, indicating that silicon incorporation was not unique to the letter samples. SNL subsequently compared the characteristics of the spore coat in the letter samples to a surrogate fermentation sample prepared at Dugway using Leighton-Doi media (Table 4-4). More than half of the spores analyzed from both the Leahy and

PHYSICAL AND CHEMICAL ANALYSES 85

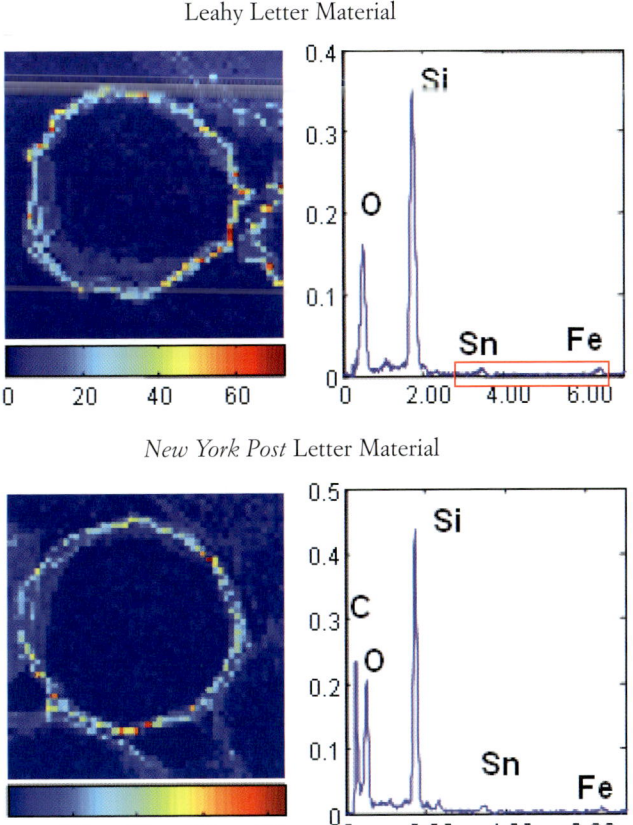

FIGURE 4-1 SEM of Leahy and *New York Post* powders.
SOURCE: SNL. This figure is the work of Sandia National Laboratories, taken or made during the course of an employee's official duties. As a work of the U.S. federal government, the image is in the public domain.

New York Post letters contained silicon in the coat. In contrast, only about one-fourth of the spores in the Dugway sample contained silicon in the coat. SNL analysis of material taken directly from RMR-1030[3] (prepared from shaken flasks at USAMRIID) showed only a few spores with silicon in the coat, and analysis of material from RMR-1029 showed no spores with silicon in the

---

[3] RMR-1030 was the label of another flask of a *B. anthracis* Ames strain spore preparation produced by and housed at USAMRIID. RMR-1030 did not resemble RMR-1029 or the letter spores in its physiochemical or genetic properties (FBI letter to the committee, December 7, 2009). The four genetic signatures (A1, A3, D, and E [see Chapters 5 and 6]) were not detected in RMR-1030.

coat. Two of the 35 Dugway plate preparations were evaluated for silicon by ICP-OES, and one showed 5 percent and the other 0.7 percent silicon. Neither of these samples showed any spores containing silicon in the coat.

Beginning in 2005, LLNL conducted a detailed study of silicon incorporation into the coats of *B. anthracis* spores. Chemical analysis was performed by nano time-of-flight secondary ion mass spectrometry (nano-SIMS), a method in which a primary beam of ions is focused onto a small region of the sample. The impact of the primary ions on the sample causes material from the sample surface to be ejected as secondary ions, which are then characterized by mass spectrometry. A chemical image is produced by scanning the primary beam across the sample. The particular strength of nano-SIMS with respect to this study was its ability to detect trace levels of silicon in the spore coat at relatively high spatial resolution. Measurements were performed on a total of 57 spore samples that included existing samples provided by the Department of Homeland Security and collaborating laboratories (Weber, 2009; FBI Documents, B1M1D7) and *B. anthracis* Sterne surrogates produced by LLNL researchers under a variety of conditions. The LLNL researchers were not given the opportunity to perform nano-SIMS on FBI evidentiary material. The goals of the Sterne experiments were to manipulate silicon levels in the spores and to quantitatively compare the amounts incorporated with previous work and predictions based on silicate solubility and precipitation.

Silicon was found in almost all of the spores examined. The high incidence of silicon detection in these experiments is most likely due to the enhanced sensitivity of nano-SIMS (impurity detection at ppm level in submicrometer regions) over SEM-EDX (Weber, 2009). The amount of silicon incorporated in the coat varied by over two orders of magnitude. While silicon incorporation tended to increase with the silicate concentration in the growth medium, a much stronger correlation was found with the amount of iron in solution. These observations and the insight they give into the mechanism of silicon uptake are the subject of a peer-reviewed publication by the LLNL team (Weber, 2009). However, the amount of silicon incorporated in these experiments did not match the letter samples. Most preparations had a silicon content below 0.1 percent per spore, with 0.3 percent being the highest amount detected—much lower than the amounts reported by SEM-EDX analysis of the letter samples, which were on the order of 1 to 2 percent per spore.

Recently, Hirota and colleagues (2010) published a study of silicon incorporation in *B. cereus* spore coats in which they found silicon in spores grown in a culture containing silicate, but none in spores grown in the absence of silicate and subsequently exposed to it. These results suggest that silicon becomes incorporated into the mother cell and then accumulates in the spore during maturation. The researchers also found that the amount incorporated was strain specific—YH64 took up 15 times more than its nearest relative NBRC15305—and that silicon incorporation enhances acid resistance.

4.5.4 Summary of the Silicon Analysis

The substantial effort devoted to the characterization of silicon in *Bacillus* spore coats resulted in new fundamental insight into microbial processes and the development of new or enhanced analytical measurement technology. (Table 4-4 presents a summary of the analytical results.) Elemental analysis of the letter samples showed that (1) the silicon content was high, (2) most of the silicon was incorporated in the spore coat, (3) the majority of spores in the samples contained silicon in the coat, and (4) no silicon was detected in the form of a dispersant in the exosporium.

The bulk silicon content in the Leahy letter could be completely explained by the amount of silicon incorporated in the spores during growth. (Not enough material was available to make this comparison for the Daschle letter.) In contrast, the *New York Post* letter had significant bulk silicon content, far exceeding that contained in the spores.

No studies have considered the effect of the chemical form of silicon (e.g., silicate impurity versus polydimethylsiloxane antifoam agent) on uptake. The inability of laboratory experiments to reproduce the silicon characteristics of the letter samples is not surprising given the complexity of the uptake mechanism.

A few spores analyzed from RMR-1030 contained silicon in the coat, but none of the spores analyzed from RMR-1029 contained silicon in the coat. Therefore, the letter samples could not have been taken directly from the flasks—a separate growth preparation would have been required.

The material in the Daschle and Leahy letters was reported to have "a high level of purity" and to have electrostatic properties that caused it to disperse readily upon opening of the letters. These properties should be regarded as qualitative observations since they were not based on quantitative physical measurements. The committee received testimony (Martin, 2010) stating that some Dugway preparations, particularly those utilizing lyophilization but no dispersant, gave products with similar appearance and electrostatic dispersibility as the letter samples, suggesting that these properties were not necessarily connected to an intentional effort to increase dispersibility through addition of a dispersant. Exogenous silicon and bentonite, which enhance the dispersibility of spore preparations, were not found in the Leahy and Daschle letters.

4.6 FEATURES OF BACTERIAL GROWTH CONDITIONS AND PROCESSING METHODS: DETECTION OF MEGLUMINE AND DIATRIZOATE

Bacterial endospores such as those produced by *B. anthracis* achieve their long-term dormancy and most of their resistance properties via relative dehydration of the spore core or cytoplasm. This dehydration results in the spores having a high density relative to the other components that remain following cultivation. A common spore purification method takes advantage of this difference in density (Tamir and Gilvarg, 1966) by centrifugation through a solution of intermediate density. The culture debris remains on top, while the high-density spores form a pellet at the bottom of the tube and can be resuspended and washed. Generally, several rounds of centrifugation and resuspension in fresh water are sufficient to remove most of the chemicals used to produce the density gradient. The most commonly used density-gradient compound is a mixture of diatrizoate and meglumine (sold under the trade name RenoCal). The presence of diatrizoate and/or meglumine in evidence samples could provide leads to their source.

In fall 2006, scientists at the FBI developed assays for the detection of diatrizoate and meglumine using liquid chromatography mass spectrometry (MS) with electrospray ionization methods. Using purified standard compounds, FBI scientists developed four-tiered assays for the detection of each compound (FBI Documents, B1M12D4).

Iohexol and metrizamide, chemicals similar to diatrizoate and also used for spore purification, were subjected to the MS analysis. These chemicals did not produce a false diatrizoate signal. The ions produced from these compounds were not described, and it is therefore not clear whether the assays would have detected them had they been used to prepare the letter spores.

Detection of diatrizoate and meglumine in purified spore samples was first demonstrated with control samples of *B. cereus* spores (FBI Documents, B1M12). The spores were purified using the standard method for spore purification at USAMRIID, involving centrifugation through two RenoCal-76 gradients, followed by two water washes. They were then gamma-irradiated in the same manner as the attack letter samples. The spores were dried and extracted with pure water and then removed by centrifugation, and the supernatant was assayed for the presence of diatrizoate and meglumine. Both compounds were clearly detectable in the control *B. cereus* spores purified using RenoCal-76.

Samples taken directly from RMR-1029 tested positive for meglumine in assay tiers 1-4 and for diatrizoate in assay tiers 3 and 4. Samples from the Leahy and *New York Post* letters tested negative for both meglumine and diatrizoate in all tiers of both assays.

This analysis was sufficient to answer one question: Were the letter spores taken directly from RMR-1029 and used without further purification? The presence of diatrizoate and meglumine in the RMR-1029 samples and their absence in the letter samples is consistent with the idea that the letter spores were not derived directly from RMR-1029. These data, in conjunction with evidence discussed elsewhere in this report (primarily concerning silicon in the spore coat), led the FBI to conclude that the letter spores were produced by a further cultivation of a sample from RMR-1029. The resulting spores were either not purified using diatrizoate and meglumine or they were washed extensively following purification, lowering the diatrizoate and meglumine to levels below the assays' limits of detection (0.01 µg/ml for diatrizoate and 0.001 µg/ml for meglumine) (FBI Documents, B1M12D4).

4.7 MEDIA COMPONENT ANALYSIS

Spores can be grown either in suspension or on agar. The FBI contracted scientists at the University of Maryland, Battelle Memorial Institute, and Pacific Northwest National Laboratory to develop new methods for determining whether the composition and source of the growth medium for spores could be ascertained. Analyses centered on detection of trace amounts of agar or blood agar, a specialized form of agar. (For a recent review of methods, see Wahl et al., 2010.)

Results of efforts to detect agar in the Leahy and *New York Post* samples through the presence of products resulting from cleavage of agar coupled to a variety of mass spectrometry techniques were inconclusive (Fenselau, 2005; Wunschel et al., 2008; Wahl et al., 2010).

Blood agar is sometimes employed as a rich medium for growth of certain microorganisms. Nutrients in blood provide a superior growth medium for some microbes. When nutrients become depleted, the microbes go into a spore-forming stage. Heme is present in blood cells and is released into the medium during spore formation, due primarily to hemolysis. Trace amounts of heme in a spore sample would indicate that blood agar was used for spore growth. Matrix-assisted laser desorption/ionization time-of-flight mass spectrometry (MALDI-TOF-MS) measurements of heme (Whiteaker et al., 2004) showed clear and specific MS signals in *Bacillus* spore samples grown on blood agar but no signal on those prepared otherwise. The protocol was able to identify heme in some FBI samples from USAMRIID with the exception of an irradiated *B. anthracis* Sterne grown on blood agar. This result may indicate that spore irradiation compromises the reliability of results from this method.

In a validation study (FBI Documents, B1M10), the agar and blood agar analyses were highly sample dependent and too sensitive to experimental conditions for the FBI to draw any conclusions from these studies. While the studies described here suggest that, under some conditions, both agar and blood agar can be detected at trace levels in spores, the information gleaned from these studies was not helpful in leading to sources for the spores used in the letter attacks. Thus the study of agar and blood agar in the evidentiary material was inconclusive and not pursued further by the FBI.

4.8 VOLATILE ORGANIC COMPOUNDS

Volatile organic compounds (VOCs) are organic compounds that have a high vapor pressure and low water solubility. They are products of microbial activity, and they are also sometimes used in the preparation of spores. They are often emitted as gases from certain solids or liquids, including common products such as paint, cleaning supplies, pesticides, and office equipment. The FBI conducted analyses to detect VOCs in letter samples that may indicate methodology used to prepare the spores (FBI Documents, B1M7D2).

The FBI used headspace gas chromatography-mass spectrometry (GC-MS) and infrared spectroscopy to detect VOCs. Headspace GC-MS was performed only on the Leahy letter sample. Trace amounts of ethanol, acetone, and t-butanol were identified. Distributions of VOCs were slightly different for Leahy and surrogate cultures, but the FBI placed little significance on the results owing to the many potential sources of these compounds.

4.9 DETERMINING WHEN THE MATERIAL WAS PRODUCED: RADIOCARBON DATING OF *B. ANTHRACIS* SAMPLES

Blinded samples for radiocarbon analysis (^{14}C) were treated with standard protocols and sent to the LLNL Center for Accelerator Mass Spectrometry (CAMS) and to the National Ocean Sciences AMS Facility (NOSAMS) at Woods Hole Oceanographic Institution during winter and spring 2002.

The radiocarbon compositions of the samples were reported as Δ^{14}C in ‰ (parts per thousand) relative to the standard date of 1950. Positive values are the result of either the rising (pre-1963 but post-1950) or the falling portion of the ^{14}C calibration curve (see Appendix A). Negative values indicate samples prepared before 1950. Seven samples were analyzed by NOSAMS and three sets of samples were analyzed by CAMS (Set #1, 20 samples; Set #2, 1 sample; and Set #3, 5 samples). Assigning actual calendar dates to specific samples is complicated by the possible incorporation of different amounts of fossil fuel carbon (with essentially no ^{14}C content) in the sample (potentially from organic solvents used in the preparation of the sample or fossil fuel burning).

CAMS assigned dates assuming a pristine environment removed from sources of fossil fuel carbon while NOSAMS added assumptions of varying amounts of fossil carbon. The analyses indicated that the Leahy sample was produced between 1998 and 2001. Radiocarbon analysis was not performed on the other letter samples.

4.10 STABLE ISOTOPE ANALYSIS

4.10.1 *B. anthracis*

Over the course of the anthrax mailings investigation, the FBI submitted a series of samples to the Stable Isotope Ratio Facility for Environmental Research (SIRFER) at the University of Utah for analysis of the stable isotope ratios of hydrogen (^{2}H/^{1}H), carbon (^{13}C/^{12}C), nitrogen (^{15}N/^{14}N), and oxygen (^{18}O/^{16}O) using its Finnegan MAT isotope ratio mass spectrometer. These

PHYSICAL AND CHEMICAL ANALYSES 91

included a variety of samples from various locations in Fort Detrick consisting of spore cultures on agar media. The samples in this set have $\delta^{18}O$ values ranging from 11.6‰ to 14.9‰ and δ^2H values between –95‰ and –76‰. The $\delta^{13}C$ values ranged from –23.9‰ to –15.2‰ and should reflect the isotopic composition of the growth medium (Kreuzer-Martin and Jarman, 2007). Among the samples of growth media, there are two distinct groups: samples with $\delta^{13}C$ from –19.1‰ to –15.9‰ and samples with values from –25.6‰ to –22.5‰. The triangles in Figure 4-1 represent the relationship between the $\delta^{18}O$ and δ^2H for these samples.

The Leahy letter evidentiary sample was received at SIRFER in late January 2004 and the analysis was completed in February 2004. The stable isotope results are $\delta^2H = -83\pm0.3‰$, $\delta^{13}C = -24.5\pm0.3$, $\delta^{15}N = +8.7\pm0.01$, and $\delta^{18}O = +18.2\pm0.4$. The relation between δ^2H and $\delta^{18}O$ values is shown as the diamond in Figure 4-2. It is apparent that the high $\delta^{18}O$ value is inconsistent with the rest of the spore samples analyzed. In an attempt to understand the rela-

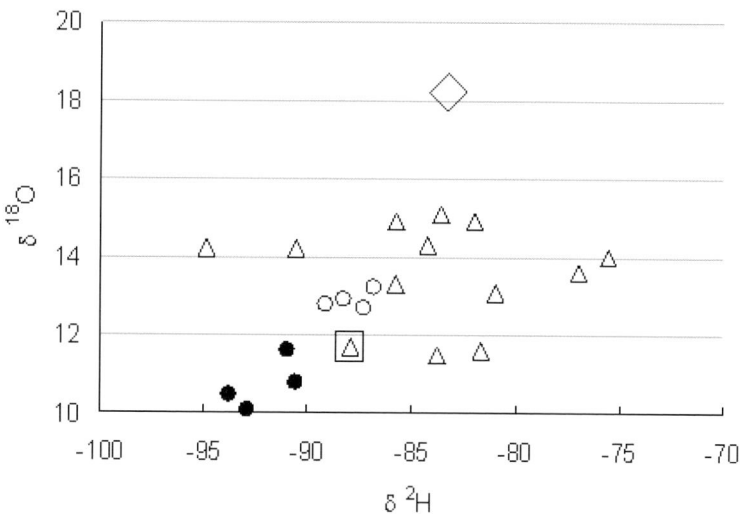

△ Ames spores   ● N samples   ○ S samples   ◇ Leahy   □ RMR 1029

FIGURE 4-2 Stable isotope results ^{18}O versus ^2H.
Stable isotopic data were obtained from a number of samples analyzed by the Stable Isotope Ratio Facility for Environmental Research (SIRFER). The triangles are from Ames spore samples obtained from various locations in Fort Detrick. The samples labeled S (open circles) and N (filled circles), received in November 2003, have different $\delta^{13}C$ signatures of ~23.5 and ~18.6, respectively, indicating different growth media. The square is from a sample of RMR-1029 and the diamond is from the Leahy evidentiary sample.
SOURCE: Courtesy of Alice Mignerey.

tive isotopic values, Ehleringer and Kreuzer-Martin used the isotopic compositions of known spore preparations of *B. subtilis* to construct a model that related the δ^2H in the sample to that of the medium and water used (Kreuzer-Martin et al., 2005). Using the δ^2H of water from Dugway as an input ($\delta^2H = -121‰$) and comparing the results they obtained for δ^2H of the medium to that of all known media, they concluded that it was highly unlikely that Dugway water was used to prepare the sample (FBI Documents, B1M9D5).

The extensive work on the influence of the growth medium on the isotopic signature of the resultant spores suggests that the Leahy sample was not produced in a liquid medium. There should be a relationship between the δ^2H and $\delta^{18}O$ in the water used in the spore preparation (meteoric water line) (Kendall and Coplen, 2001); this constrains the possible values of $\delta^{18}O$ used in the initial spore preparation. Using data gathered for *B. subtilis* (Kreuzer-Martin et al., 2003) and other strains of *B. anthracis* (A0256 and Sterne), Ehleringer and Kreuzer-Martin concluded that data for the Leahy sample are inconsistent with the sample having been grown in a liquid medium with purified tap water. They posit that it was likely to have been grown on agar medium with significant enrichment in the $\delta^{18}O$ signature due to evaporation (FBI Documents, B1M9D14). Stable istotope analysis was not performed on the other letter samples.

Additional research on the effects of growth media and culture conditions on the resultant spore isotopic composition could develop this approach into a useful forensic tool that might provide leads in future investigations.

4.10.2 Water Samples

The FBI collected samples of tap water at 20 locations in 18 cities and sent them to SIRFER for δ^2H and $\delta^{18}O$ analysis. The correlation between the δ^2H and $\delta^{18}O$ values is very close to the meteoric water line (MWL) (Kendall and Coplen, 2001). As expected, the samples from Dugway in Utah are the lightest (most negative) with $\delta^2H = 120‰$ and $\delta^{18}O = -15‰$. Those from Miami are very close to zero (the value of mean ocean water). Three samples were taken in the Frederick area, one at USAMRIID, one from a well, and one from the city municipal supply. All gave consistent values around $-45‰$ for δ^2H and $-7.5‰$ for $\delta^{18}O$. These results proved useful in providing starting water isotopic ratios as input for the interpretation of the spore sample isotopic ratios presented in Section 4.10.1 (FBI Documents, B1M9D1).

4.10.3 The Envelope Measurements

SIRFER conducted stable isotope analysis for δ^2H, $\delta^{13}C$, and $\delta^{18}O$ on four sets of envelopes, each with three samples: one untreated, one subjected to the decontamination procedure of irradiation, and one treated with ninhydrin. The whole envelope material was measured, as was the cellulose extracted from the envelope material. There was no statistical difference in any of the untreated samples for $\delta^{13}C$, which had an average value of −23.7‰. Comparison with the irradiated samples showed no measurable change, but the ninhydrin treatment decreased the $\delta^{13}C$ by about 0.66‰. There was a small difference in the $\delta^{18}O$ between the samples designated as "O" and those as "N", with the "O" samples 29.3±0.3‰ compared to 30.4±0.1‰ for the "N" samples. The FBI drew no conclusions based on these results (FBI Documents, B1M9D9).

4.11 COMMITTEE FINDINGS

Finding 4.1: The committee finds no scientific basis on which to accurately estimate the amount of time or the specific skill set needed to prepare the spore material contained in the letters. The time might vary from as little as 2 to 3 days to as much as several months. Given uncertainty about the methods used for preparation of the spore material, the committee could reach no significant conclusions regarding the skill set of the perpetrator.

The DOJ *Amerithrax Investigative Summary* indicates that, because of the extraordinary high spore powder concentrations and the exceptional purity of the material in the Washington, D.C., letters, "the anthrax mailer must have possessed significant technical skill" (USDOJ, 2010, p. 14). The FBI did not present a definite theory on how and when propagation, purification, and drying took place, nor on what specific skills would be required to perform these tasks. Nonetheless, inferences made by the FBI concerning the time, skill, and equipment required for spore preparation were said to be significant considerations in their narrowing of the list of potential suspects (USDOJ, 2010, pp. 29-33, 36-38). In discussions with the FBI at the January 2011 meeting, the FBI informed the committee that some of its consulting experts referred to the letter preparations as being of "vaccine quality", which narrowed the list of potential suspects but that they investigated any and all individuals without regard to their specific skill sets. The FBI further stated that there were too many variables to be able to quantify the time or specify the equipment used to prepare the letter materials. However, FBI officials indicated that inferences about required skills or time for spore preparation were never the sole criterion for eliminating suspects (FBI/USDOJ, 2011).

There are distinct professional judgments of the time that this work would require, with some estimates as low as 2 or 3 days and others in the range of many months. The distinctions are based on different judgments of the time required for propagation, purification, and drying, among other variables, as well as the state of the starting material available to the individual(s) in question. In particular, it is not known whether some of the initial steps might have occurred well in advance of the letter attacks. The committee cannot resolve these distinctions because a particular production method or the steps involved in production were not identified.

Finding 4.2: The physicochemical methods used primarily by outside contractors early in the investigation were conducted properly.

Electron microscopy, aerosol particle size analysis, and bulk and spatially resolved elemental compositions were appropriate first steps for characterizing the samples. Follow-up experiments to determine the spatial distribution of silicon in letter material and the mechanism of silicon uptake also were appropriate, as were efforts to search for possible elemental and molecular signatures. While recognizing the circumstances of the time, and the urgency of these studies, the committee notes that the physical science investigations were not pursued to the extent that they could resolve important discrepancies.

The FBI plan for the chemical and physical characterization of samples (FBI Documents, B1M1D5) contains a flow diagram defining the approach and refers to the key scientific methods to be used. The committee, however, was not shown any record of how the FBI made decisions about which experiments to pursue or which approaches to abandon. The committee also did not see records of detailed discussions about which samples were to be studied with each of the characterization techniques (see Table 3-2 in Chapter 3 for information the committee was able to gather from all of the submitted materials).

Finding 4.3: Although significant amounts of silicon were found in the powders from the *New York Post*, Daschle, and Leahy letters, no silicon was detected on the outside surface of spores where a dispersant would reside. Instead, significant amounts of silicon were detected within the spore coats of some samples. The bulk silicon content in the Leahy letter matched the silicon content per spore measured by different techniques. For the *New York Post* letter, however, there was a substantial difference between the amount of silicon measured in bulk and that measured in individual spores. No compelling explanation for this difference was provided to the committee.

ICP-OES analysis indicated a silicon content of the bulk *New York Post* letter material of 10 percent by mass, while SEM-EDX performed by SNL demonstrated silicon in individual spore coats at a level corresponding to 1 percent by mass per spore. At the January 2011 meeting, the FBI attributed this difference to a limited amount of sample available (only one replicate was performed for ICP-OES analysis) and the heterogeneous character of the *New York Post* letter. An explanation based on the heterogeneous character implies that the specific samples analyzed were not representative of the letter material. In such a case, additional samples should have been analyzed to determine representativeness. If such data exist, they were not provided to the committee. Lacking

PHYSICAL AND CHEMICAL ANALYSES   95

this information, one cannot rule out the intentional addition of a silicon-based substance to the *New York Post* letter, in a failed attempt to enhance dispersion. The committee notes that powders with dispersion characterization similar to the letter material could be produced without the addition of a dispersant.

Early in the investigation, AFIP performed SEM-EDX analysis of a *New York Post* letter sample and found regions in the sample having high silicon content but no oxygen, suggesting the presence of silicon-rich material that was not related to nanoparticulate silica. While this observation could have led to an explanation for the difference between the bulk and individual spore measurements, follow-up experiments apparently were not performed. The committee notes that this information was not made available to it or to the FBI until spring 2010.

Finding 4.4: Surrogate preparations of *B. anthracis* did reproduce physical characteristics (purity, spore concentration, dispersibility) of the letter samples, but did not reproduce the large amount of silicon found in the coats of letter sample spores.

Surrogate preparations by DPG, using *B. anthracis* from the Leahy letter as the starter source (FBI Documents, B1M13D3), reproduced the general physical characteristics of the letter samples (purity, spore concentration, dispersibility) but not the silicon chemical characteristics. Surrogate preparations showed that samples having bulk silicon content up to 5 percent could be prepared without intentional addition of silicon dispersant. However, none of the DPG surrogate preparations analyzed for silicon in the spore coat were similar to the *New York Post*, Daschle, and Leahy letter samples with respect to either the amount per spore of silicon incorporated in the coat or the fraction of spores observed to contain silicon in the coat. Furthermore, the committee sought, but could not obtain, a detailed explanation of the thought process that went into selection of the DPG methods or their relationship to the Buran and Abshire preparations. The committee acknowledges that there were many more possible scenarios for spore preparation than could have been feasibly explored with available resources and in a reasonable period of time. However, it was not clear to the committee how the subset of surrogate preparation methods was selected and whether these choices were based on an understanding informed by the investigation or on other assumptions about the approach taken to produce the evidentiary materials.

Finding 4.5: Radiocarbon dating of the Leahy letter material indicates that it was produced after 1998.

The spores in the letter were not taken directly from a stockpile produced many years ago. One or more recent growth steps would have been required, although it is not possible to pinpoint the time frame for that growth. Comparing the hydrogen isotope ratios from water from DPG to a model for all known media, Kreuzer-Martin and colleagues concluded that it was highly unlikely that water from DPG was used to prepare the sample; however, it was not possible to identify the location where the spores were prepared.

Finding 4.6: The flask designated RMR-1029 was not the immediate, most proximate source of the letter material. If the letter material did in fact derive from RMR-1029, then one or more separate growth steps, using seed material from RMR-1029 followed by purification, would have been necessary. Furthermore, the evidentiary material in the New York letters had physical properties that were distinct from those of the material in the Washington, D.C. letters.

SEM-EDX measurements showed no silicon in the coats of spores taken directly from RMR-1029, whereas the majority of spores analyzed from the *New York Post*, Daschle, and Leahy letter materials contained silicon in the coat. Based on recent studies of the mechanism of silicon incorporation, silicon could have been incorporated in the coats of the letter spores only if spores from RMR-1029 were subjected to one or more subsequent growth steps. Another observation consistent with a separate growth step was the detection of *B. subtilis* in the *New York Post* and Brokaw letter material but not in RMR-1029 (discussed in Chapter 5). The detection of meglumine and diatrizoate in RMR-1029 but not in the Leahy and *New York Post* samples also is consistent with this finding; however, it is not conclusive because it might have been possible to rinse these impurities away without requiring later growth. Some of these findings, as well as others, indicate that the New York letter materials were prepared separately from the materials in the Washington, D.C., letters. The presence of *B. subtilis* in the New York but not the Washington letter materials and the different physical properties of the materials indicate that the two sets of letter materials were prepared separately.

5

Microbiological and Genetic Analyses of Material in the Letters

5.1 INTRODUCTION

As discussed in Chapter 2, the bacterium *Bacillus anthracis* (*B. anthracis*) is the causative agent of the disease anthrax. Anthrax generally affects grazing mammals (e.g., cattle, sheep, horses). Anthrax in humans, particularly inhalational anthrax, is rare and occurs mainly in individuals with occupations involving the handling of hides, hair, or bone from infected animals. Inhalational anthrax can also be a sign of a biological attack with *B. anthracis* spores, especially in individuals without likely exposure to infected animals or their products. This was the case in 2001, when a number of cases of both cutaneous and inhalational anthrax were diagnosed among media and postal employees and others after the delivery of letters containing suspicious powders in several places.

In October 2001, clinical reporting of human anthrax cases spurred a broad epidemiological investigation to identify the source of the illnesses as well as any other infected parties (see Chapter 3 for a timeline and a more detailed discussion of this epidemiological investigation). Identification of *B. anthracis* as the cause of the 22 cases of illness and five deaths in 2001 was determined by clinical laboratory means (CDC, 2001a, b, c). Because these illnesses and deaths appeared to have resulted from a bioterrorist attack, immediate efforts were undertaken to identify a common source for the outbreak, including molecular genetic analysis of the causative agent.

5.2 IDENTIFICATION OF THE *B. ANTHRACIS* STRAIN

The first step in the search for a source of the anthrax-causing powders in the 2001 mailings was to identify which "strain" (or strains) of *B. anthracis* was used in the attack. Disparate cases of human anthrax in Connecticut, New York, Florida, and Washington, D.C., and contaminated environmental locations in the last three of these sites were all linked when a single *B. anthracis* strain, the Ames strain, was identified in all of these cases and associated locations (Keim et al., 2008). As noted in Chapter 2, *B. anthracis* is one of the most genetically homogeneous microorganisms known (Keim et al., 2000). Nonetheless, even in the most homogeneous species there are usually some differences in genome sequences among populations. These sequence differences, although few in number, are sufficient to characterize subgroups, or "strains." Strains are members of the same species, but their differences reflect the divergence of sublineages as they evolved over time (Keim et al., 2000). Among *B. anthracis* populations, a variety of strains had already been recognized even before sequencing technology enabled detailed characterizations of genome differences among strains.

Work performed by Paul Keim and others well before the 2001 anthrax attacks had resulted in the development of several molecular methods to detect genetic differences among *Bacillus* species as well as among isolates of *B. anthracis*. In the mid-1990s, work by Hendersen and colleagues (1995) and Anderson and colleagues (1996) led to the identification of a 12-nucleotide variable number tandem repeat (VNTR) sequence (called *vrr*A for "variable repeat region A") that provided the first molecular marker that distinguished among *B. anthracis* strains. The basis of this marker was shown to be differences in the number of repeated sections of this genetic sequence, and five different variations were detected. Subsequently, VNTR analysis at multiple genetic loci (Multiple Locus VNTR Analysis or MLVA) enabled the characterization of 426 *B. anthracis* isolates with 89 distinct genotypes (Keim et al., 2000).

Another approach, amplified fragment length polymorphism (AFLP) analysis, has been particularly useful for examining differences between *B. anthracis* and close relatives, such as *B. cereus* and *B. thuringiensis* (Keim et al., 1999, 2008; Hoffmaster et al., 2002; Keim et al., 2008). The AFLP technique had been used to identify about 30 variable regions and provided an ability to profile portions of the genome of a large number of diverse *B. anthracis* strains. In addition, the *pagA* gene, which encodes the protective antigen (PA) protein (one of the three anthrax toxin proteins discussed in Chapter 2) also had been sequenced (Price et al., 1999). Because of the importance of *pagA* in the development of immunity to anthrax, this gene was of interest in determining whether a particular strain might have been genetically altered, or "engineered," for increased effectiveness as a weapon (Hoffmaster et al., 2002).

These new molecular approaches, combined with the creation of a collection of strains from many of the world's geographic regions, greatly enhanced scientific capabilities for identifying genetic variations among anthrax strains at that time. Using these methods, Keim and colleagues (1999, 2000) had established a picture of the evolutionary lineages of *B. anthracis*. These methods for rapid, reliable molecular subtyping were also critical in determining the identity of the clinical and environmental isolates in the 2001 anthrax attacks. Although a complete genome sequence provides the most effective genetic signature for identifying an organism, at the time of the attack mailings no *B. anthracis* genomes had yet been completely sequenced and published (Keim et al., 2008).

Using MLVA at eight loci identified by Keim and his colleagues, scientists from the Centers for Disease Control and Prevention's (CDC's) Laboratory Response Network subtyped 135 *B. anthracis* isolates (samples) collected from the attack victims, letter powders, and environmental samples, and determined that all these isolates were likely to have been derived from a common source (Hoffmaster et al., 2002). In addition, CDC scientists sequenced the *pagA* genes from a subset of these isolates and concluded that none of the isolates appeared to have been engineered (Hoffmaster et al., 2002). (The use of additional tests performed under the aegis of the FBI to assess the possibility of genetic engineering is discussed below.) All attack-associated isolates were identified as MLVA genotype 62 and PA genotype I. Genotype 62 is the genotype of the Ames strain commonly used worldwide for laboratory research for vaccine development. The PA I genotype was also identical to the Ames strain PA genotype. These results led CDC to conclude that the *B. anthracis* strain used in the attacks was indistinguishable from the Ames strain (Hoffmaster et al., 2002).

As early as October 2001, samples from the spore-laden envelopes and clinical isolates (including one from Robert Stevens, the deceased Florida patient and index case) were also sent to Paul Keim's laboratory at Northern Arizona University. The Keim laboratory had already established the *B. anthracis* MLVA sequence database that contained information on more than 1,000 samples from around the world. This database proved useful for identifying the *B. anthracis* strain in the forensic samples.

Beginning in January 2002, Keim also began conducting genetic testing at the request of the FBI on isolates of *B. anthracis* provided by the United States Army Medical Research Institute for Infectious Diseases (USAMRIID), which had received evidentiary samples from the FBI. The first 18 evidentiary samples (designated Batch E0001) received by the Keim laboratory were handled according to FBI chain of custody requirements and were initially not identified. An initial MLVA-8 analysis found that all but two of the samples ("Connecticut samples") provided by USAMRIID were identical to and consistent with the Ames strain genotype (Keim, 2002a). An expanded analysis of 15 MLVA loci (Keim, 2002b) yielded similar results, with all but 3 forensic samples matching the Ames strain. One sample differed from the Ames strain in that it had lost the pXO2 plasmid (see Chapter 2), but was otherwise identical to the others. Keim noted that plasmids are commonly lost during culture and that this loss may have occurred prior to shipment of the sample to his laboratory, but was not to be interpreted as necessarily indicating that the original forensic sample was pXO2 negative. The two Connecticut samples proved to be distinct from the other evidence, but were identical to each other. Their genotypes matched 10 isolates from China in the Keim laboratory database that did not have a genotype or strain designation at that time, but most closely resembled reference strain Genotype 61. At the time, Keim noted that these samples clearly did not match the Ames strain. These samples proved to be from a separate case (Malakoff, 2002) and not from the elderly Connecticut patient who died from inhalational anthrax. The FBI was initially interested in these samples until it was determined that they were not related to the anthrax letter investigation. The Keim laboratory continued to receive and strain type the *B. anthracis* samples submitted to the FBI Repository (see Chapter 6) over the next six years.

5.3 WAS THE *B. ANTHRACIS* IN THE LETTERS GENETICALLY ENGINEERED?

An important investigative issue was whether the *B. anthracis* strain used in the mailings had been genetically engineered. For example, the FBI was interested in whether antibiotic resistance genes or virulence factors had been introduced into the attack strain's genome from other strains or species, and whether there were any other mutations in the sample, engineered or otherwise, that might help investigators determine its source.

To address some of these questions, Paul Jackson and colleagues at the Los Alamos National Laboratory (LANL) analyzed samples from the attack envelopes from October 2001 through mid-2002. The materials tested included samples of the progenitor Ames strain isolated from the dead cow in 1981 (Ames "Ancestor"); an isolate from the deceased Florida patient, Robert Stevens; isolates from the Brokaw, *New York Post*, and Daschle letters; and several stocks from USAMRIID, including a sample from flask RMR-1029. The samples were tested for possible indications of genetic engineering using DNA sequencing (see Box 5-1) and polymerase chain reaction (PCR) amplification (see Box 5-2) to look for (1) the presence of genes encoding resistance against the antibiotics ciprofloxacin, tetracycline, erythromycin, bleomycin, kanamycin, and chloramphenicol; (2) modifications of the protective antigen, edema factor, and lethal factor protein genes; and (3) inserted sequences derived either from cloning vectors (plasmids) known from the literature to have been used to engineer *B. anthracis* or from the insertion of the cereolysin genes of *B. cereus,* reported (Pomerantsev et al., 1997) to have conferred upon the strain an ability to evade protective immunity induced by some anthrax vaccines (FBI Documents, B1M4D2). The LANL scientists reported that "none of the isolates assayed showed any indication of genetic manipulation based on the presence of markers normally associated with genetic manipulation of *B. anthracis*."

It should be noted that the LANL investigators did not look for other less obvious alterations that also might have indicated that the organisms in the evidentiary samples had been genetically engineered. Indeed, they acknowledged that a well-financed laboratory could exploit or develop other cloning vectors and other methods for genetic manipulation without leaving clear molecular

BOX 5-1
Genome Sequencing

Until the 1970s, DNA was the most difficult molecule in biology to analyze because of its enormous length and its "monotonous" chemical structure. The DNA molecule is a string of chemical building blocks—the nucleotides or "bases" adenine (A), thymine (T), cytosine (C), and guanine (G). DNA sequencing is the process of determining the exact order of these building blocks in a piece of DNA. For example, "ATCGGCTAA" is part of a DNA "sequence." Today, DNA sequencing has become indispensable for basic research and for numerous applications, such as disease diagnostics, biotechnology, systematics, and forensic biology. The earliest DNA sequencing methods were developed in the 1970s and were laborious and very slow. But the Human Genome Project, which began in 1990 and was largely completed in 2003, greatly stimulated the development of new sequencing technologies. These are largely automated and are orders of magnitude faster than earlier efforts. For example, in 2001 it could take about a year to sequence the genome of a bacterium like *B. anthracis*, whereas today such a process requires only a few hours.

Genome sequencing generally requires that the genome being studied first be broken into smaller pieces. This process is usually carried out by enzymes that "cut" the DNA into short fragments. In an alternate approach called "shotgun" sequencing, the DNA of interest is mechanically broken into random overlapping fragments. Each fragment is sequenced numerous times, and the genome is reassembled using computational methods to order the fragments based on the regions of overlap. If there is sufficient overlap, the genome sequence can be considered complete or "closed." In bacteria, which usually have circular chromosomes, this means that the entire circle of the sequence is known. This technique works particularly well for small genomes, such as those of bacterial species that do not have extensive regions of repetitive nucleotide sequences. For the much larger genomes of animals and plants, the DNA of interest may also be broken into pieces and then cloned by inserting the fragments into bacteria, which make copies of the DNA when they divide and reproduce. The cloned fragments are then mapped against the genome being sequenced. The mapping reduces the likelihood that regions containing repetitive sequences (much more common in eukaryotic genomes) will be assembled incorrectly.

The enormous increases in speed and efficiency of DNA sequencing have led to a revolution in scientists' ability to identify mutations quickly and precisely. Recently, the term "deep sequencing" was coined to describe this approach of simultaneous sequencing of massive numbers of short fragments derived from a mixture of genomes, such as might exist in an evolving population derived from a single cell that has, over time, accumulated genetic variants. These millions of short sequences are then ordered by computer programs, enabling the identification of single nucleotide polymorphisms (SNPs) and other genetic variants.

> **BOX 5-2**
> **The Polymerase Chain Reaction Technique**
>
> The PCR technique provides a rapid means to amplify (increase the number of copies of) DNA segments of interest. Knowledge of the DNA sequence to be amplified is used to design two specific, but fairly short, synthetic DNA strands, or oligonucleotides, that are complementary to the sequence of DNA to be amplified. These oligonucleotides serve as "primers" for DNA synthesis and determine the segment of DNA amplified. In PCR, the original double-stranded DNA is first heated to separate the strands. The separated strands are then cooled in the presence of an excess of the two primers, which hybridize with the complementary sequences in the strands of DNA being studied. The mixture is then incubated with the nucleotides that are the raw materials for DNA synthesis and an enzyme called "DNA polymerase." This enzyme synthesizes new DNA starting from the primers and copying the DNA strand to which the primer is bound. The result of the first cycle of PCR produces two new double-stranded DNA molecules that each contains one strand of the original DNA and one strand from the primer. This cycle of denaturation, hybridization, and synthesis is then repeated many times, increasing the number of copies of the DNA sequence of interest. Each cycle generates fresh templates for further amplification, and only the sequences bracketed by the primers are amplified, while regions that lack priming sites are not. After a number of these cycles, a substantial proportion of the reaction mixture corresponds to the amplified DNA.
>
> SOURCE: Alberts, B., Johnson, A., Lewis, J., Raff, M., Roberts, K., and P. Walter. (2002). *Molecular Biology of the Cell,* 4th ed. New York: Garland Science.

signatures, or at least the signatures that the investigators sought. However, the subsequent completion of the genomic sequences of the Ames Ancestor and later of the letter isolates (see Section 5.5.5 below) strongly supported the findings of the LANL group (see also Box 5-1 on genome sequencing). As noted by Read and colleagues (2002), identification of genes that have been altered or inserted deliberatively in potential bioweapons agents is facilitated by complete genome sequencing. No further testing related to the issue of genetic engineering of the attack powders was performed after mid-2002 aside from the genome sequencing.

Prior to the 2001 letter attacks, the Institute for Genomic Research (TIGR) had begun sequencing the genome of the "Porton Down" isolate of *B. anthracis* (Read et al., 2002). The genomes of many bacteria have multiple parts, typically including a single large circular chromosome and one or more smaller plasmids. In particular, most *B. anthracis* strains carry two plasmids, called pXO1 and pXO2, that encode proteins required for virulence but that are not essential for the bacteria to grow under laboratory conditions. However, the Porton isolate had been rendered avirulent by "curing" (eliminating) both plasmids from the Ames strain (Read et al., 2003), making it a less than optimal choice for a reference genome. Shortly after the letter attacks, the National Science Foundation provided critical funding that allowed TIGR to sequence to draft quality the isolate from the spinal fluid of the deceased Florida patient, Robert Stevens (Ames "Florida"). Here, "draft quality" refers to a genome sequence for which a small fraction of the nucleotide bases remains uncertain, there are remaining small gaps in the coverage of the complete genome, or both. This work was described in a paper by Read and colleagues in 2002, and included a comparison of the Florida and Porton isolates and an examination of a group of isolates of the Ames strain obtained from various laboratories prior to 2001 as well as an isolate from a Texas goat obtained in 1997. The latter was the only other isolate of the Ames strain known to have been collected in the field aside from the 1981 dead cow "Ancestor" isolate. In this initial set of comparisons, 11 sequence differences were found between the chromosomes of the Porton and Florida isolates (Read et al., 2002); however, thesel differences were also found in all the other Ames isolates tested. From these data, and the understanding that the Porton Down strain was derived from earlier isolates at USAMRIID (Read et al., 2002), it was inferred that the mutations in the Porton strain occurred after the 1982 transfer of the strain to Porton Down. No sequence differences appeared to distinguish the isolate of the Florida victim from many of the isolates of the Ames strain present in various laboratories before the attack.

In spring 2003 TIGR also initiated the sequencing of the Ames Ancestor and completed this work in October 2003 (Ravel, 2010). Unlike the Florida isolate, which was sequenced only to draft quality, the Ames Ancestor sequence was "closed," that is, assembled into one contiguous sequence. Annotation and analysis of the Ames Ancestor sequence continued until mid-2004 and it was released to GenBank on June 1, 2004. (The paper announcing the Ames Ancestor sequence was not, however, published until 2009; see Ravel et al., 2009). The Ames Ancestor sequence served as the high-quality reference genome needed for the comparative genomics work that TIGR later performed on colony morphotypes identified from the attack letter materials (see below).

The apparent absence of chromosomal differences between the attack strain and Ames strains had important implications, both positive and negative, for the investigation. On the positive side, the findings strongly supported the inference that the attack strain had come—directly or indirectly—from a laboratory that possessed the Ames strain. Also on the positive side, the findings supported the conclusion that the attack strain had not been engineered to make it resistant to treatment or more virulent. On the negative side, the absence of distinctive sequences in the attack strain seemed to mean that it would not be possible to use genetic markers to trace the attack material to one particular source among the various institutions (and laboratories within the institutions) that possessed the Ames strain.

Nonetheless, an important clue as to the source of the strain in the letters came from microbiologists at USAMRIID who discovered that the attack material contained at low to moderate frequency several subtypes of *B. anthracis* that produced morphologically distinct colonies. These colony morphological variants (or "morphotypes") retained their distinctive appearance when single cells from the colonies were regrown into new colonies. This persistence meant that the morphotypes were genetically distinct from the standard (wild-type) Ames strain in the samples; that is, they apparently contained a mutation or mutations that caused them to produce their distinctive-looking colonies. That several morphotypes could be distinguished indicated that different morphotypes contained different mutations. The morphotypes appeared to be spontaneous mutants that arose during the preparation of the batches of spores that were eventually used in the attacks.

As described and discussed later in this chapter and in Chapter 6, TIGR also sequenced the genomes of several of these morphotypes to identify the mutations that distinguished them from the wild-type Ames strain and from each another (FBI Documents, B1M5D1-2). These mutations played a central role in the investigation (USAMRIID, 2005; FBI Documents, B1M2D12; Worsham, 2009), helping the FBI to trace attack materials to a possible source.

The investigation summaries provided to the committee in December 2010 refer to the presence of a pE03 vector referred to as an 'Israeli cloning vector' among certain repository isolates (B3D1). On the January 11, 2011 meeting the FBI was asked to clarify what was known about the vector. The committee was told that this vector was a derivative of the commonly used cloning vector pBR322 that was found in some isolates, and that it had no forensic value to the investigation (FBI/USDOJ, 2011).

5.4 *B. SUBTILIS* CONTAMINATION OF THE NEW YORK SAMPLES

A finding that was initially of high forensic interest was the discovery, based on cultivation techniques, that the powder from the letters sent on September 18 to two New York City addresses (the *New York Post* and Tom Brokaw at NBC) contained a mixture of *B. anthracis* and, at a frequency of about 1 to 5 percent, a non-*B. anthracis* bacterium. The contaminating bacterial species was identified by the CDC as *B. subtilis* on the basis of 16S rRNA gene sequencing and later whole genome sequencing (CDC, 2001a; FBI Documents, B2M1D2). *B. subtilis* is a ubiquitous bacterial species that is readily isolated from environmental samples from around the world. The identification of the contaminant as *B. subtilis* was at first considered of potential importance because certain strains of *B. subtilis* are widely used in academic and industrial laboratories. Hence, if the contaminant had proved to be a particular laboratory strain, it might have provided a clue to the origin of the New York City powders.

Whole genome sequencing analysis carried out and reported in 2004 by TIGR of an isolate referred to as GB22 from the *New York Post* letter showed high (98 percent) similarity, but not identity, to the published sequence of the standard laboratory strain *B. subtilis* 168 (Kunst et al., 1997). In 2006, TIGR developed 95 PCR assays for 23 *B. subtilis* loci in the evidentiary sample that differed from the reference (*B. subtilis* 168) genome. The amplified DNA regions were compared using gel electrophoresis. (DNA sequencing of these amplified regions would have been a more definitive approach.) The *B. subtilis* isolates from the *New York Post* and Brokaw letters were identical to each other at all 23 loci, indicating that they were the same strain. (The whole genome sequence of the *B. subtilis* from the Brokaw letter was not determined, however, so their identity was not definitively demonstrated.)

Because the 95 PCR assays would have been cumbersome to perform on larger collections of samples, the FBI Laboratory next identified four genetic markers in the GB22 letter strain, three of which (designated ID 65, ID 91, and ID 107 by TIGR) were rare in a survey of 72 *B. subtilis* strains isolated from around the world. These strains were obtained from the NRRL (formerly the Northern Regional Research Laboratory) collection of the U.S. Department of Agriculture and the American Type Culture Collection, and they were meant to represent a geographically and genetically diverse collection (FBI Documents, B2M4D2). The three rare markers distinguished the GB22 strain from the other strains. The fourth marker (*sboA*) was common to all the *B. subtilis* strains examined. This combination of markers was designed first to determine whether any *B. subtilis* was present in additional samples based on the presence of the *sboA* marker and, second, to determine whether such samples contained a strain that might be similar to or the same as the GB22 strain from the New York letters.

The FBI Laboratory developed TaqMan (see Box 5-3) real-time PCR (RT-PCR) assays for the four markers, and these assays were provided to the Naval Medical Research Center (NMRC) and the National Bioforensic Analysis Center (NBFAC) at the Department of Homeland Security's National Biodefense Analysis and Countermeasures Center, where the assays were validated by blind testing. NMRC and NBFAC used the assays to evaluate over 300 evidentiary samples. Only two *B. subtilis* strains from these samples were found that matched GB22 at all four loci. But when TIGR followed up by further characterizing these two samples using the complete set of 95 PCR assays, they proved to be genetically different from GB22 (FBI, 2009). NBFAC later (2007) also tested all Ames strain samples in the FBI Repository (see Chapter 6) for the presence of *B. subtilis* contamination (see NBFAC analytical result reports, November 2006-December 2007, FBI Documents, B2M4D3-15). Although 322 out of 1057 repository samples tested positive for the *sboA* nucleic acid sequence, further testing showed that none of these 322 samples was positive for the rare ID 65 marker in GB22 (NBFAC, 2007; FBI Documents,

> **BOX 5-3**
> **The TaqMan Technique**
>
> The TaqMan® technique is highly sensitive (Easterday et al., 2005a) and reliable as a diagnostic tool (Easterday et al., 2005b), and allows the detection of genetic differences between samples even at the level of SNPs (Van Ert et al., 2007b). It uses an oligonucleotide probe that anneals to both the wild-type and mutant target DNA. The probe is labeled with both a fluorescent tag and a fluorescence quencher and binds tightly to the exact complementary sequence in the target DNA. PCR is initiated using primers that anneal nearby. One of these primers is designed such that it anneals only to a template containing the specific allele to be detected. Two primer sets are generally used, one that is specific for the wild-type allele and a second that is specific for the mutant allele. If the primer anneals and Taq polymerase synthesizes a new strand along the template, then the bound fluorescent TaqMan probe will be digested by the exonuclease activity of the advancing polymerase, thereby releasing the fluorophore and producing a signal that indicates the presence of the particular allele. If the primer does not anneal and Taq polymerase does not synthesize DNA, then the oligonucleotide probe will remain bound and intact, and the fluorescent tag will not emit detectable fluorescence due to the close proximity of the quencher.

B2M4D13). (If any samples had been positive for the presence of the ID 65 marker, analyses for the ID 91 and ID 107 markers would have also been performed.) In short, many repository samples were contaminated with *B. subtilis*, although apparently not by the same strain as in the *New York Post* letter. Ultimately, the FBI concluded that the testing for *B. subtilis* did not provide useful information leading to the source of the New York letter materials—GB22 is apparently an environmental strain of unknown origin that could not be traced to any particular source.

5.5 IDENTIFICATION AND CHARACTERIZATION OF COLONY MORPHOLOGICAL VARIANTS IN THE EVIDENTIARY MATERIAL

5.5.1 Why Was the FBI Interested in Colony Morphotypes?

Any microbial geneticist can attest to the fact that close scrutiny reveals unusual individual variants in a population of microbes. Such variants often carry genetic alterations that produce noticeable phenotypic changes in physiology, behavior, or morphology. When the variants are observed at the level of the physical appearance of colonies as they grow on agar plates, those variants are often referred to as "morphotypes." As the selective pressures for rapid growth under laboratory conditions may differ from those experienced by organisms in their natural environment, genetic variants (mutants) may arise that replicate faster and, over time, replace the "wild-type" strain during repeated cycles of laboratory culture. This process represents, in effect, rapid evolution leading to a population becoming better adapted to its particular laboratory conditions.

Analyses of the spore samples from the attack letters from as early as November 2001 (FBI Documents, B1M2D7) revealed the presence of colony morphotypes whose stability suggested that they resulted from the presence of genetically distinct subpopulations. The specific set of genetic alterations in a population might provide a useful profile for that population and if it can be demonstrated to be present in two different sample populations, might suggest that the two were derived from a common source. With sufficient knowledge about the identities of such genetic variations and the frequencies with which they arise in the population under specific culture conditions, the statistical significance of the similarities between the two populations might, in principle, be calculated. Under some circumstances, the chance of two *independent* populations containing the same genetic variant subpopulations might be so small that it could be concluded with high confidence that they were derived from a single source. With this goal, the FBI pursued detailed characterization of the phenotypic and genetic variation among the evidence samples.

5.5.2 Background Information on Morphotypes

Given the rapid generation times (many generations per day) and large populations (often billions of cells) typically observed in laboratory bacterial cultures, it is highly likely that genetic variants will arise in cultured populations. If a genetic variant in a population is able to initiate growth more quickly, grow more rapidly, or sustain growth longer as conditions become less favorable, it tends to increase its frequency in the population. This process can lead to cultures with multiple genetically distinct subpopulations. While this basic process of selection drives evolution in nature, it can present problems for genetic studies in the laboratory, as the characteristics of an organism may change over the time frame of a scientific study (Elena and Lenski, 2003).

Microbiologists usually seek to avoid this phenomenon using two primary methods. First, stock cultures are stored under nongrowth conditions in either a dried or frozen state, since most mutations do not arise in the absence of growth. Fresh samples of that nonvarying stock are then used to initiate each new set of analyses. Second, cultures retrieved from the stock are streaked on growth medium in such a way that individual colonies are obtained and each colony contains a population derived from a single cell (i.e., a clonal population). Some genetic variants that make up a minor proportion of the stock population may be recognizable based on their atypical colony morphology, and those variants can be avoided in the generation of cultures for further uses.

It is well known in laboratories that study bacterial spore formation (sporulation) that low-frequency sporulating (oligosporogenous) variants appear in a bacterial population at a readily observable frequency (Velicer et al., 1998; Michel et al., 1968). Furthermore, laboratory selective pressures can result in these variants becoming dominant members of the population and supplanting the sporulation-proficient parent strain. Two phenomena may increase the abundance of asporogenous (nonsporulating) or oligosporogenous (sporulating with reduced frequency) variants above that of other types of variants. When a population encounters a decline in nutrients, sporulation-proficient cells cease growth and enter the dormant spore state. Sporulation-deficient variants may continue growth to some degree at the expense of their neighbors, increasing their own relative abundance in the population. If this aged population of cells and spores is eventually transferred to fresh medium, the nonsporulated cells may also resume growth more rapidly than the spores, again increasing their abundance in the resulting population. These may be selective factors leading to the frequent observation of oligosporogenous variants in laboratory cultures.

Studies of *B. anthracis* phenotypic variants, many of which were shown to be oligosporogenous, were previously carried out by Patricia Worsham and Michele Sowers at USAMRIID (Worsham and Sowers, 1999) as part of an effort to identify useful strains for vaccine research. Working with multiple strains (e.g., Ames, Sterne, Vollum, and others) of *B. anthracis*, Worsham and Sowers identified numerous phenotypic variants (Worsham and Sowers, 1999). Their initial screen for phenotypic variation was growth on medium containing a dye called Congo red. The dye accumulated by wild-type *B. anthracis* colonies results in the colonies assuming a salmon-pink color. Poorly sporulating colonies do not accumulate the dye, resulting in a white appearance that can be easily identified by eye. Further study of the white colony morphotypes revealed that they were sporulation deficient. While there is no clear physiological explanation for the coincidence of the absence of Congo red binding and sporulation deficiencies, no such explanation is required for the practical use of this morphotype screening method. One of these phenotypic variants was carefully characterized as carrying a mutation in *spo0A*, a primary regulatory gene for entry into sporulation (Hoch, 2000). Indeed, mutants of *spo0A* and other genes governing entry into sporulation were historically recognized in experiments in which cultures of cells sporulating in rich medium were maintained for prolonged periods of time (Michel et al., 1968).

Reinforcing the view that laboratory conditions frequently enrich for mutants defective in sporulation, a recent publication by Sastalla and colleagues (2010) reports that repeated passage *of B. anthracis* on rich medium results in the accumulation of sporulation mutants, which were recognized by their distinctive colony phenotype. Many of the mutants contained mutations in the sporulation regulatory gene *spo0A*, as had been observed by Patricia

Worsham (Worsham and Sowers, 1999). These included point mutations as well as deletions and insertions. Sastalla and colleagues further showed that the *spo0A* mutants exhibited a shorter lag time than did the wild type when inoculated on rich medium. Thus, one factor contributing to the enrichment of sporulation-deficient mutants upon repeated passage is that at least some sporulation mutants resume rapid (exponential phase) growth more quickly than does the wild type.

5.5.3 Detection and Characterization of Morphotypes in the Anthrax Letter Samples

Part of the analysis of the anthrax letter evidence samples involved the preparation of relatively low-density liquid suspensions of spores recovered from the letters and the spreading of these suspensions on the surface of a solid growth medium. This procedure allowed each viable spore to germinate, grow, and divide in isolation, producing a clonal population of millions of cells—all derived from a single cell—in the form of a colony. This is a standard microbiological practice used for the enumeration of viable spores in a sample. The method also allows the visual detection of variants in the population that may carry genetically heritable traits responsible for altered phenotypic properties, although the sensitivity and discriminatory capacity of this method are low. Relatively early in the investigation (fall 2001), Terry Abshire, a microbiologist at USAMRIID, observed the presence of multiple phenotypic variants (morphotypes) among the colonies generated from the anthrax letter spore samples. The analysis of these variants eventually became a major feature in the FBI's effort to identify the most likely source of the specific culture used to produce the letter samples. To the advantage of the FBI, Patricia Worsham was one of the USAMRIID scientists that assisted with the analysis of the letter spore samples. Due to her extensive experience with *B. anthracis* and her specific experience studying *B. anthracis* phenotypic variants, Worsham and her colleagues were readily able to apply methods for the study and characterization of phenotypic variants in the letter samples.

In early analysis of the Leahy letter sample, spores were plated on sheep blood agar and incubated for three days. On these plates, Abshire observed the presence of morphotypes with more defined colony edges and a yellow tint relative to the majority of the colonies. Based on this observation, Worsham was asked to develop a protocol to characterize morphotypes in the multiple evidentiary samples. The goal of this analysis was the identification of phenotypic and eventually genetic signatures of the letter samples that might contribute to determining the source of the letter material.

Spore samples from the Daschle, Leahy, and *New York Post* letters were suspended in water at a low density, such that they could be spread to give rise to between 10 and 50 colonies per plate, allowing the observation of individual colony phenotypes. Thousands of colonies from each sample were visually inspected. The initial plating was on sheep blood agar plates to allow the most efficient colony recovery. The plates were incubated at either 26°C or 37°C with room air, or at 37°C with an atmosphere containing 20 percent CO_2. These conditions were employed to allow the detection of variants that might be apparent only under particular growth conditions. While the characteristics of colony morphotypes are most obvious following incubation for 48 hours or more after nutrient exhaustion and sporulation, these starvation conditions were avoided in order to diminish the likelihood of the selection of new variants that were not present in the initial evidentiary samples. Plates were inspected after 18 to 22 hours of incubation, which was sufficient to allow the development of sizable colonies. Most classes of colony morphotypes were identified predominantly by the presence of a yellow tint (although the committee notes that this tint is not apparent in Figures 5-1, 5-2, and 5-3, while one class exhibited a smaller, more opaque colony appearance.

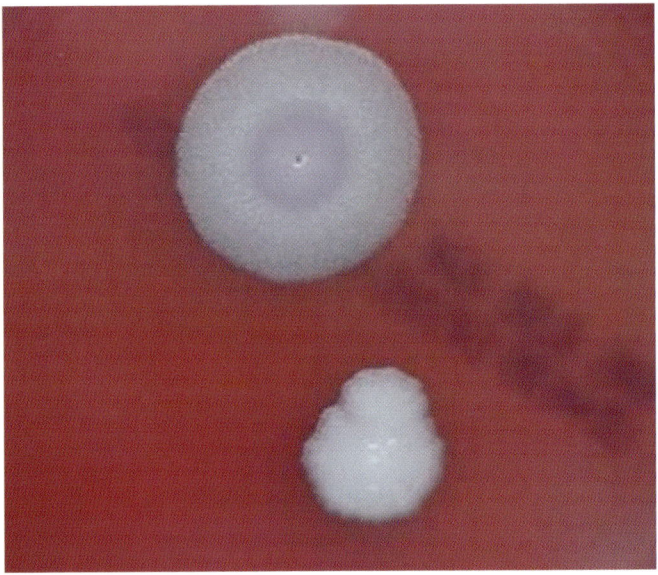

FIGURE 5-1 *B. anthracis* Colony Morphotype "A." Photograph of colonies formed by growth of *B. anthracis* cells on blood agar. The colony on the top displays the morphology designated "Type A." The colony on the bottom displays the typical wild-type morphology.
SOURCE: USAMRIID. This image is a work of the United States Army Medical Research Institute for Infectious Diseases, taken or made during the course of an employee's official duties. As a work of the U.S. federal government, the image is in the public domain.

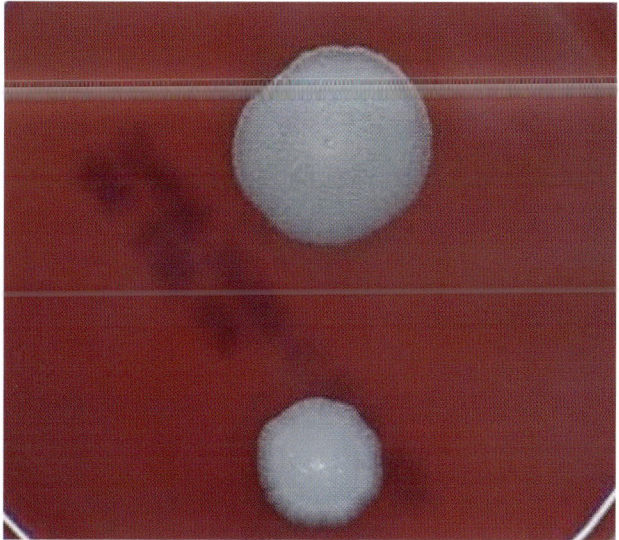

FIGURE 5-2 *B. anthracis* Colony Morphotype "B." Photograph of colonies formed by growth of *B. anthracis* cells on blood agar. The colony on the top displays the morphology designated "Type B." The colony on the bottom displays the typical wild-type morphology.
SOURCE: USAMRIID. This image is a work of the United States Army Medical Research Institute for Infectious Diseases, taken or made during the course of an employee's official duties. As a work of the U.S. federal government, the image is in the public domain.

Upon the identification of each candidate variant colony, the colony was scraped from the culture plate. To produce an archival stock of the variant, 50 percent of the recovered material was stored at −70°C. This is a standard microbiological practice for the maintenance of strains under conditions that prevent the further accumulation of genetic variation. The remainder of the recovered material was again streaked onto a fresh sheep blood agar plate and incubated at 37°C for 18 to 22 hours in order to produce numerous colonies for each candidate morphotype. This new growth was inspected and compared to a nonvariant isolate to determine whether the phenotype of each candidate morphotype was stable over multiple generations, indicating the presence of an underlying genetic change (mutation), and whether the recovered material produced a uniform colony appearance, indicating a pure (clonal) culture of the morphotype. Colonies were collected from this second plate to produce an additional stock at −70°C that could be used for multiple analyses without sacrifice of the archival stock.

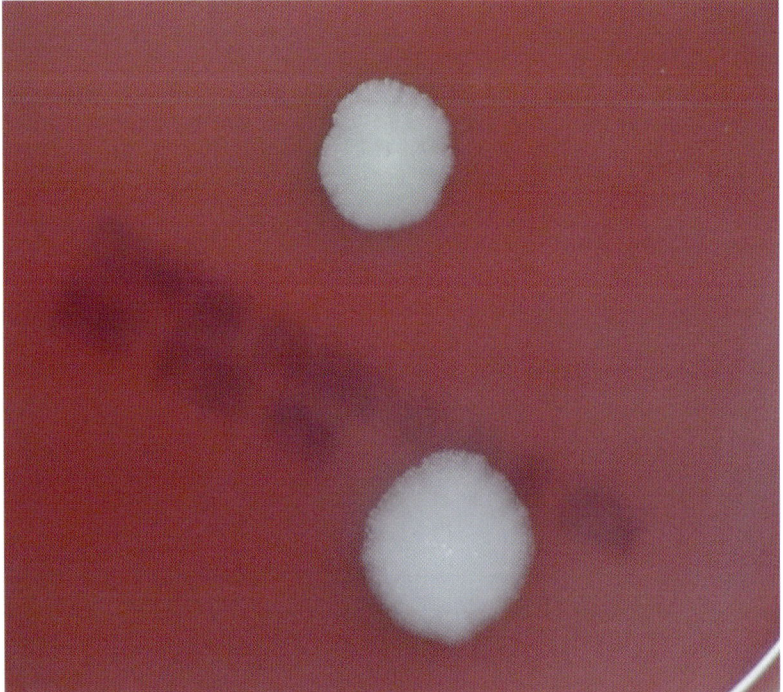

FIGURE 5-3 *B. anthracis* Colony Morphotype "E." Photograph of colonies formed by growth of *B. anthracis* cells on blood agar. The colony on the top displays the morphology designated "Type E." The colony on the bottom displays the typical wild-type morphology.
SOURCE: USAMRIID. This image is a work of the United States Army Medical Research Institute for Infectious Diseases, taken or made during the course of an employee's official duties. As a work of the U.S. federal government, the image is in the public domain.

Each morphotypic variant recovered was subjected to numerous phenotypic tests that had been used in earlier studies (Worsham and Sowers, 1999) to differentiate variants from the wild-type strain. In each case, candidate variants were observed side by side, on the same plate of medium, with colonies of the wild-type strain as well as isolates of several previously identified morphotype classes. Variants were examined for Congo red binding, degree of colony spreading, degree of colony pigmentation, and "bull's-eye" appearance of the colony (which might be due to variation between concentric rings in colony thickness, pigmentation, or other physiological traits). Sporulation was scored by microscopic observation of material recovered from colonies on two different media:

TABLE 5-1 Phenotypic Characteristics of the Morphotypes

	Colony size	Colony spread	Pigmentation	Hemolysis
Morphotype A	Large	Yes	Bull's eye	++
Morphotype B	Large	Yes	Altered	++
Morphotype C/D	Large	Yes	Bull's eye	+
Morphotype E	Small	No	Opaque	−

sheep blood agar, which is a rich medium not conducive to efficient sporulation, and Leighton-Doi medium, a minimal medium that is normally used for spore production. Hemolysis, the rupturing of red blood cells and alteration of the red heme color, can be detected after growth on sheep blood agar for 48 hours at 37°C followed by refrigeration for two weeks, during which the hemolytic reaction becomes more visibly pronounced.

Following extensive testing, four predominant morphotype classes were identified in the Leahy letter material, although there were also other less abundant types. The four were named A, B, C/D (two initial similar classes, C and D, were eventually merged), and E (also called opaque) (see Table 5-1). All of the morphotypes were oligosporogenous, producing significantly fewer spores than the wild type on all media tested. Morphotypes A, B, and C/D exhibited large zones of hemolysis that were not produced by the wild-type strain. The defining characteristic of morphotype A was a large spreading colony with a bull's-eye center. Morphotype B was similar to A in colony size but did not produce the bull's eye and produced slightly different pigmentation on some media. Morphotype C/D was also similar in appearance to A, but produced a smaller zone of hemolysis and was somewhat more proficient at spore production. Morphotype E was unique in producing colonies that were smaller than those of the wild type, had more regular edges, and were more opaque. Morphotype E also exhibited temperature-sensitive sporulation, producing significantly more spores at 26°C than at 37°C, yet still fewer than the wild type at either temperature. Screening of samples from the Daschle and *New York Post* letters revealed morphotype classes that were phenotypically similar to those from the Leahy letter. (Due to a scarcity of material from the Brokaw letter, no screening of samples for colony morphotypes was performed on that material [Hassell, 2010].)

5.5.4 Selection Criteria for Genetic Variations Used in Screening

Classes of morphotypes present in the letter samples had been characterized by the scientists at USAMRIID as early as January 2002, and the FBI began to consider methods for screening the FBI Repository (FBIR) (see Chapter 6) of *B. anthracis* isolates to identify a source culture. Phenotypic screening of the

FBIR samples was deemed an impractical method for several reasons. First, phenotypic screens of the types described above are relatively slow, labor intensive, and highly dependent on the trained eyes of the investigator to identify variant colonies. Second, similar phenotypic variations can be associated with different genetic alterations located in either the same or widely separated genetic loci. The presence of similar colony morphotypes in two samples would not provide direct genetic evidence to link the two sample populations. Third, phenotypic screens are insensitive and do not reliably detect rare variants. Identification of the specific mutations associated with each phenotypic variation was required for the development of definitive assays to detect the presence of shared mutations in multiple strains within the repository. Such DNA-based assays are rapid, sensitive, compatible with high-throughput methods, and definitive to the level of nucleotide sequence.

Scientists selected representative morphotype isolates as well as control wild-type isolates from each of three letter samples for detailed genetic analysis. Several criteria were used for this selection. First, the scientists needed to be able to distinguish the variants from the wild-type colonies on plates. Second, these particular morphotypes must have been present at a high enough frequency for the scientists to identify them repeatedly. The third essential criterion was the apparent presence of the morphotype in each of the three letter samples (Leahy, Daschle, *New York Post*) that were subjected to this analysis. The final selection of morphotypes focused on four variants: A, B, C/D, and E. There were other morphotypes found in the letter materials, but they were not used for further forensic testing. Worsham and colleagues at USAMRIID quantified the percentages of variants by randomly picking about 370 isolated colonies from plates made using dilutions from the Leahy letter. These colonies were 79 percent wild-type morphology, 6.7 percent C/D morphotype, 1.1 percent B morphotype, 1.3 percent A morphotype, and 4.9 percent E morphotype (other morphotypes accounted for the remaining fraction) (FBI Documents, B1M2D12). It is important to note that two identical-looking morphotypes need not, and often did not, have the same genotype. Indeed, as discussed in the next section, two independent isolates exhibiting similar colony morphotypes might have mutations in different genes or even different mutations in the same gene. Also, some colonies identified as morphotypic variants may not have had any mutation, as the distinction between genetic and nongenetic variation is not always clear. Thus, it was crucial to identify differences in nucleotide sequence as an unambiguous signature of different mutant subpopulations.

5.5.5 Whole Genome Sequencing of Morphotype Isolates

To determine whether the genetic alterations associated with each colony morphotype might be suitable for use as forensic markers, the genome sequences of multiple morphotype isolates were determined (Table 5-2). Genomic DNA

TABLE 5-2 *B. anthracis* Isolates Analyzed by the Institute for Genomic Research (TIGR)

TIGR ID	FBI ID	Origin	Morphotype	Sequencing status
GBA	Ames Porton	Porton Down	Wild type	12X – closed
GB6	Ames Ancestor	Texas/USAMRIID	Wild type	12X – closed
GB8	LL10/E3	Leahy letter	A	8X – closed
GB9	LL9/E2	Leahy letter	B	8X
GB10	LL1/E1	Leahy letter	Wild type	8X
GB11	PL10/E6	*New York Post* letter	A	8X
GB12	PL9/E7	*New York Post* letter	B	8X – closed
GB13	PL1/E9	*New York Post* letter	Wild type	8X – closed
GB15	DL10/E4	Daschle letter	A	NS
GB16	DL9/E5	Daschle letter	B	NS
GB17	DL1/E8	Daschle letter	Wild type	NS
GB18	LL6/E10	Leahy letter	C	12X
GB19	LL7/E11	Leahy letter	D	NS
GB23	LL18	Leahy letter	E	12X
GB24	LL19	Leahy letter	E	NS

NS = not sequenced
SOURCE: FBI Documents, B1M5D1-2.

extracted from the colony morphotypes identified by USAMRIID was provided by Paul Keim to TIGR, where it was prepared for genome sequence analysis. Plasmid libraries were produced from the genomic DNAs and shotgun sequencing was carried out to produce approximately 8X (in some cases 12X) average coverage of the genome. However, most of these genome sequences were not closed (i.e., assembled into one contiguous sequence). In some cases (e.g., the samples from the Daschle letter), no sequencing was performed. In these cases the TIGR scientists used PCR to test whether the same genetic differences in the sequenced samples were also present in the unsequenced ones (FBI Documents, B1M5D1). Again, as previously noted, efforts were not undertaken to identify these or other morphotypes in the Brokaw letter.

Control wild-type isolates from each letter were found to possess no genetic differences from the Ames Ancestor strain (FBI Documents, B1M5D1-2). The genome sequences of each of the chosen morphotype isolates did, however, exhibit differences from the wild-type isolates in particular genetic regions of the *B. anthracis* chromosome or, in the case of the E morphotype, in the pXO1 plasmid. Some samples of morphotype A contained a single nucleotide polymorphism (SNP) while others carried a duplication. SNPs were also found for morphotypes B and C, while D contained a chromosomal deletion and E had a deletion in the pXO1 plasmid. Following the identification of these sequence differences, unsequenced morphotype isolates were further tested for the presence of the same sequence differences using PCR amplification and sequencing of the amplified DNA. The results are summarized in Table 5-3. These data revealed that multiple isolates of some of the morphotypes (e.g., B) were associated with a single genetic change while others (e.g., isolates of the A morphotype) exhibited several different sequence variations in the same chromosomal region.

The genotypes of the A morphotype isolates from three letters (Leahy, *New York Post*, and Daschle) were of two general kinds. The first was a SNP

TABLE 5-3 Further Genetic Characterization of the Morphotype Isolates

Morphotype class	Affected locus	Type of mutation	Genotype examined in greater detail	Letter source	Assay method developed
A	One copy of 16S rRNA gene	Insertions in different sites overlapping a 16S rRNA gene	A1, 2024 bp	Leahy[a] Daschle[b] New York Post[b]	qPCR
			A2, 2608 bp	New York Post[a]	Not used
			A3, 823 bp	Leahy[b] Daschle[a] New York Post[b]	qPCR
B	spo0F promoter	SNP in intergenic region	B	Leahy[a] New York Post[a]	Not used
C	Sensor *his* kinase	SNP producing stop codon	C	Leahy[a]	Not used
D	Sensor *his* kinase	258 bp deletion	D	Leahy[a]	Taqman + PCR
E	Putative response regulator (plasmid)	9 or 21 bp deletion, or SNP	E, 9 bp deletion	Leahy[a] Daschle[c] New York Post[c]	Taqman + PCR

NOTE: bp = base pair; qPCR = quantitative polymerase chain reaction; SNP = single nucleotide polymorphism.
  [a] Source of original morphotype isolate with this genotype.
  [b] Sample subsequently found to contain DNA carrying this genotype.
  [c] Source of subsequent morphotype isolate demonstrated to have this genotype.
SOURCE: FBI Documents, B1M5D1-2.

in a gene encoding a K+ uptake protein. This SNP was found only in the *New York Post* letter material, but not in the Porton Down, Ames Ancestor, or Leahy genomes, and no further forensic use was made of this mutation. The second kind of A morphotype genotype involved large insertions in one of the eleven copies of the *B. anthracis* 16S rRNA gene. A 2024 bp (base pair) insertion (later termed the A1 genotype) and an 823 bp insertion (A3 genotype) were both found in the materials from the Leahy, Daschle, and *New York Post* letters, and both were subsequently chosen for the development of two separate assays to be used to screen the FBI Repository of Ames strain samples (see Chapter 6). A 2608 bp insertion (A2 genotype) was found only in the *New York Post* letter and an assay was developed for the detection of this genotype by Commonwealth Biotechnologies, Inc., (CBI) in Richmond, Virginia. In validation testing, the performance of the A2 assay was surpassed by that of the A1 and A3 assays. Consequently, the FBI decided not to use the A2 assay for evidence analyses. Eventually, many more variants ranging from 822 to 2608 bp were found among other samples provided to TIGR (FBI Documents, B1M5D4).

B morphotype isolates from the Leahy and *New York Post* were fully sequenced (Table 5-2), while the morphotype isolate from the Daschle letter was studied using PCR. The sequencing revealed that the Leahy and *New York Post* letters contained an identical SNP in the noncoding region between the *spo0F* gene and an adjacent gene, and this SNP was not present in the Ames Ancestor or Ames Porton reference samples. PCR amplification and resequencing of the amplicon confirmed the presence of the same SNP in the same region of the Daschle morphotype B isolate. The SNP was the replacement of a thymine (T) with a cytosine (C). The two open-reading frames are divergently transcribed, so this intergenic region likely contains the promoters for these genes, one of which (*spo0F*) plays a key role in governing entry into sporulation. This mutation may explain why the morphotype is sporulation deficient. The gene expression patterns of these mutants were not examined because these kinds of experiments were considered outside the scope of the investigation. The consistent presence of the B SNP provided a second potential genotypic signature for comparing FBIR samples to the letter materials, but it was not used by the FBI to screen repository samples because multiple efforts by contractors to develop assays for this SNP failed (see below), nor were any other SNPs representing single base pair mutations used to that end (see Chapter 6).

The morphotype C and D isolates shared a very similar phenotype. For the C morphotype, only material from the GB18 sample from the Leahy letter was sequenced (Table 5-2). Again, the TIGR team's reasoning was that other samples did not need to be sequenced because any polymorphisms found in GB18 could be tested by PCR later. The GB18 C morphotype analysis found one SNP corresponding to a nucleotide change from G to A, creating a stop codon in a histidine sensor kinase ("*his* kinase") gene, a member of a family of proteins that regulate gene expression. When the D morphotype sample

GB19 from the Leahy letter was subsequently examined for the presence of the C SNP in the same gene sequence, a 258 bp deletion was found instead. This genetic deletion resulted, in turn, in a deletion of 86 amino acids in the same *his* kinase protein. The SNP found in GB18 is located in the chromosomal region that is deleted in GB19. Thus the similarity in the C and D phenotypes could be explained since both the C SNP and the D deletion likely produced a nonfunctioning protein from the same *his* kinase gene, which plays a role in sporulation. The *New York Post* and Daschle letters were not tested for the C and D morphotypes.

The morphotype E isolate from the Leahy letter (GB23) was sequenced and had no chromosomal mutations. Instead, this strain had a 21 bp deletion in the pXO1 plasmid. The deletion was located in a gene encoding a putative gene expression regulator. A PCR/resequencing assay was used to test for the presence of the same deletion in the GB24 Leahy letter sample. This sample contained a 9 bp deletion in the same genetic locus. The PCR/resequencing analysis was run on a series of other blind samples relevant to the investigation according to TIGR's 2005 report (FBI Documents, B1M5D4). Some of these additional strains carried the 9 or 21 bp deletions, but others contained a SNP representing a single point mutation (CGT → TGT) in the same locus that appeared to create defects in the corresponding proteins severe enough to interfere with normal function, although it was beyond the scope of TIGR's work to test this.

Table 5-4 provides a summary of the distribution among the case letters of the morph genotypes that were ultimately used for screening of the FBIR.

TIGR completed this stage of the study by analyzing a set of samples (listed in Table 7, FBI Documents, B1M5D4) that included additional evidentiary samples as well as noncase strains from the Keim laboratory's scientific collections. TIGR tested these samples using the assays developed for the various genotypes. These additional analyses showed that only the colony morphotype samples themselves contained the specific polymorphisms identified by the TIGR team, which was interpreted to mean that these genotypes represented

TABLE 5-4 Distribution Among the Anthrax Letters of the Genotypes Selected for FBIR Screening

Letter	Genotype A1	Genotype A3	Genotype D	Genotype E
Leahy	+	+	+	+
Daschle	+	+	NT	+
New York Post	+	+	NT	+
Brokaw	NT	NT	NT	NT

NT = not tested
SOURCE: Hassell (2010).

unique markers suitable for further forensic use. The A1, A3, D, and E genotypes were employed for the development of validated assays that were used to screen samples of the Ames strain collected by the FBI from all domestic and foreign sources that it was able to identify. The details of this screening are provided in Chapter 6.

5.5.6 Development and Application of Assays for the Genotypes

Genotypes A1 and A3

CBI was the contractor selected by the FBI to develop the genetic assays for the A1 and A3 morphotypes and test the FBI Repository samples (Chapter 6). CBI began work for the FBI in mid-2002 (FBI Documents, B2M5D2). The A morphotypes that were analyzed most thoroughly were found to contain large insertions overlapping a 16S rRNA gene (FBI Documents, B1M5D1). Although the insertion was of a different size in each of the three A morphotypes, all three had insertions in the same locus. The A1 genotype had a 2024 bp insertion and was originally identified in an isolate from the Leahy letter. During assay development by CBI this allele was also detected in DNA derived from bulk Daschle and *New York Post* letter spores (Hassell, 2010). The A3 genotype contained an 823 bp insertion that was originally identified in an isolate from the Daschle letter, but during assay development by CBI this allele was also detected in DNA derived from bulk Leahy and *New York Post* letter spores (Hassell, 2010). The A2 genotype had a 2608 bp insertion and was originally identified in an isolate from the *New York Post* letter. No acceptable assay for A2 was developed, and it is not known whether the A2 allele was present in spore material from the other letters. The assays developed by CBI used the TaqMan analytical technique (Didenko, 2001), which is an adaptation of PCR (see Boxes 5-2 and 5-3).

CBI completed its validation studies in February 2004. Limits of detection were estimated at 0.005 percent for the A1 genotype assay and 0.001 percent for the A3 assay in a background of 20 nanograms (ng) of Ames Ancestor DNA. Appropriate reaction controls were also developed. Sequencing of amplicons served as a final confirmatory step. The A1 and A3 genotype assays were chosen by the FBI for use in analyzing the FBIR samples (see Chapter 6) as problems (e.g., high number of false positives) with the validation of the assay for genotype A2 ultimately caused this assay to be abandoned. It should be noted that assay development and validation took almost two years.

Genotypes B and D

Three companies—CBI, IIT Research Institute (IITRI), and Midwest Research Institute (MRI)—were hired to develop assays for the B and D genotypes. None of the contractors was successful in developing a reliable B assay. In addition, the FBI expressed a preference not to use assays directed at SNPs (FBI, 2009). Consequently, the FBIR samples were not screened for the B genotype. The IITRI and MRI assays for the D genotype were both accepted by the FBI and used to screen the FBIR samples. The D genotype had a 258 bp deletion and was originally identified in an isolate from the Leahy letter. It was never determined whether this allele was present in the spore populations in the other evidentiary letter samples. Assay development and validation in each case took almost one year.

IITRI assay (FBI Documents, B2M7): A technical proposal for assay development for the D genotype was submitted by IITRI in July 2004 and validation of this assay was completed in April 2005. Using TaqMan/PCR (Boxes 5-2 and 5-3) this assay detected the genotype D when it was present at levels as low as 0.01 percent relative to the Ames Ancestor background. The repository screenings using the IITRI assay for the D genotype began in May 2005 and were completed in early 2007.

MRI assay (FBI Documents, B2M8): MRI submitted its technical proposal for development of the D deletion assay to the FBI in July 2004 and assay development was completed in June 2005. It used RT-PCR and had a detection limit of 0.01 percent in the Ames Ancestor background. Three approaches were used to increase sensitivity: closely spaced primers, short annealing time (15 seconds), and confirmation of reaction amplicon with melt curve and fragment size analysis. This second D assay also went forward for screening the repository and other samples. Screening began in December 2005 and was completed in October 2007.

Genotype E

The E morphotype was identified from the Leahy, Daschle, and *New York Post* letters (FBI Documents, B1M5D4). Although there were apparently several different mutations that produced the "opaque" phenotype, all appeared to involve the same gene on the pXO1 plasmid. One isolate from the Leahy letter material carried a 21 bp deletion in this gene, and another isolate also from the Leahy letter had a 9 bp deletion in the same region of that gene. Both of these deletions were also found in E isolates from the Daschle and *New York Post* letters using PCR and sequencing of this locus (Hassell, 2010). Other E isolates contained a single bp substitution in the same gene.

The 9 bp deletion was chosen for the development of an assay for use in screening the FBIR samples. The assay was developed in 2005 using TaqMan technology, validated, and applied to the repository by TIGR under contract from the FBI in 2007 (FBI Documents, B2M9). Preparations of purified mutant and wild-type DNA were mixed in amounts covering a 1,000,000-fold range of ratios of mutant-to-wild-type DNA, and the assay was shown to reliably detect the mutant genome when present at 0.01 percent of the total DNA in the sample. In addition, the assay did not produce false positives for the presence of the mutant allele using varying amounts of wild-type DNA. This assay was approved by the FBI for testing repository samples, which was performed from June to August of 2007.

5.6 COMMITTEE FINDINGS

Finding 5.1: The dominant organism found in the letters was correctly and efficiently identified as the Ames strain of *B. anthracis*. The science performed on behalf of the FBI for the purpose of *Bacillus* species and *B. anthracis* strain identification was appropriate, properly executed, and reflected the contemporary state of the art.

Finding 5.2: The initial assessment of whether the *B. anthracis* Ames strain in the letters had undergone deliberate genetic engineering or modification was timely and appropriate, though necessarily incomplete. The genome sequences of the letter isolates that became available later in the investigation strongly supported the FBI's conclusion that the attack materials had not been genetically engineered.

In the first few months following the attacks, isolates from the letters and other sources were examined only for the presence of some obvious and expected signs of genetic engineering. This examination was not exhaustive and would have missed less obvious or less well recognized signatures of deliberate genetic alteration. Had the case not involved the Ames strain of *B. anthracis*, with its relatively brief history and high degree of characterization, this limitation could have been a serious one.

Finding 5.3: A distinct *Bacillus* species, *B. subtilis*, was a minor constituent of the *New York Post* and Brokaw (New York) letters, and the strain found in these two letters was probably the same. *B. subtilis* was not present in the Daschle and Leahy letters. The FBI investigated this constituent of the New York letters and concluded, and the committee concurs, that the *B. subtilis* contaminant did not provide useful forensic information. While this contaminant did not provide useful forensic information in this case, the committee recognizes that such biological contaminants could prove to be of forensic value in future cases and should be investigated to their fullest.

Although the *B. subtilis* isolates in the two New York letters appeared to be closely related, the *B. subtilis* isolate in the Brokaw letter was not fully sequenced, and therefore the presumed identity of the two isolates was not definitively demonstrated. Although *B. subtilis* was found in several hundred repository samples, the strains in these samples did not match the isolates found in the New York letters. Biological contaminants could prove to be of great forensic value and should be investigated to their fullest in future cases.

Finding 5.4: Multiple colony morphotypes of *B. anthracis* Ames were present in the material in each of the three letters that were examined (*New York Post*, Leahy, and Daschle), and each of the phenotypic morphotypes was found to represent one or more distinct genotypes.

This important discovery greatly facilitated the subsequent laboratory investigation and is a testament to the critical importance of attentive, thoughtful scientists who were prepared to explore unexpected results in the setting of a forensic investigation.

Finding 5.5: Specific molecular assays were developed for some of the *B. anthracis* Ames genotypes (those designated A1, A3, D, and E) found in the letters. These assays provided a useful approach for assessing possible relationships among the populations of *B. anthracis* spores in the letters and in samples that were subsequently collected for the FBI Repository (see also Chapter 6). However, more could have been done to determine the performance characteristics of these assays. In addition, the assays did not measure the relative abundance of the variant morphotype mutations, which might have been valuable and could be important in future investigations.

In the course of developing the assays that were used to screen the FBIR samples for the four genotypes, procedures were employed to examine both the specificity and sensitivity of the assays, including analyses of defined mixtures of genotypes at known proportions. However, the repository included both homogeneous and heterogeneous samples, in unknown proportions, and the extent of genetic diversity in the heterogeneous samples was also unknown. More could have been done to determine the performance characteristics, including reproducibility of results, under the actual conditions associated with the repository samples.

In addition, these assays were not used to quantify the relative abundance of the genotypes in the FBIR samples and the evidentiary materials. Measurement of relative abundance of genotypes might have helped clarify the relationship between the evidentiary spore samples and whether they were derived from the same or different cultivation events.

Finding 5.6: The development and validation of the variant morphotype mutation assays took a long time and slowed the investigation. The committee

recognizes that the genomic science used to analyze the forensic markers identified in the colony morphotypes was a large-scale endeavor and required the application of emerging science and technology. Although the committee lauds and supports the effort dedicated to the development of well-validated assays and procedures, looking toward the future, these processes need to be more efficient.

Future cases may not allow for a time frame as lengthy as that of the anthrax letters investigation. Assay development and validation took almost two years in some cases, for reasons that are not clear to the committee. The committee recognizes that the experience gained in the case, as well as faster and greatly improved technologies, could help speed future investigations. These factors alone, however, may not be sufficient for all contingencies. In particular, future cases could involve less well documented or less easily grown species and strains, and precious investigation time could be lost because of the need to establish basic information about the relevant organism's biology and population genetics. In addition, original attack material (in this case, the powder in the anthrax letters) may not be available in all bioterrorism scenarios. Also, in some future cases of bioterrorism the attacks may continue until the perpetrators are identified and apprehended.

6

Comparison of the Material in the Letters with Samples in the FBI Repository

6.1 INTRODUCTION

The previous chapters have established the following main conclusions: First, the physical evidence associated with the anthrax spores in the attack letters did not reveal the source of the materials in the letters. Second, genetic analyses of the spores established that the anthrax belonged to a particular subtype known as the Ames strain. Third, while no genetic differences were observed between the predominant *Bacillus anthracis* genotype in the letters and the canonical Ames strain, FBI contractors discovered several subpopulations of *B. anthracis* cells in the letters that each produced distinct colony morphologies after growth on agar plates, and some of these subpopulations possessed specific genetic mutations that could be identified by specific sequence-based assays developed by the FBI and its contractors.

In this chapter, we describe the establishment of the FBI Repository (FBIR) of Ames strain samples, created by the FBI to look for the *B. anthracis* subpopulations found in the letters with the hope of being able to identify the source of the *B. anthracis* spores used in the attacks. The chapter also outlines the history of RMR-1029, the spore-containing flask at the U.S. Army Medical Research Institute for Infectious Diseases (USAMRIID) that became a focus of the FBI investigation, at least in part because it contained all of the subpopulation genotypes that were assayed.

The results obtained from screening the repository samples are described, including the evidence that appeared to implicate RMR-1029 as the source of the spores in the attack letters. Particular attention is given to the limitations of applying formal statistical methods to these results and of trying to quantify the strength of the relationship between the spores in the attack letters and those in RMR-1029. The committee then discusses evidence bearing on a disputed sample submitted by the suspect in this case as well as follow-up experiments performed by the FBI to determine whether that sample had come from its stated source, RMR-1029 (see *Amerithrax Investigative Summary*,

USDOJ, 2010, pp. 75-79, for a description of the circumstances surrounding the disputed sample). The chapter ends with the committee's major findings regarding the genetic evidence relevant to the source of the *B. anthracis* spores used in the letter attacks.

6.2 CREATION OF THE FBI REPOSITORY (FBIR)

At the time of the anthrax mailings, the federal government had not systematically collected information on which laboratories possessed anthrax. Although the "Select Agents" program was created in 1996 by the Antiterrorism and Effective Death Penalty Act (Public Law 104-132), this statute governed primarily the transfer of biological agents between research laboratories. The Act directed the Secretary of Health and Human Services (HHS) to issue regulations governing the transport of biological agents with the potential to pose severe threats to public health and safety through their use in bioterrorism, the so-called "select agents" (NRC, 2009). The authority to regulate select agents was delegated by the HHS Secretary to the Centers for Disease Control and Prevention (CDC). To ensure that these agents were transferred only between responsible parties, CDC required that any laboratories that might transfer select agents be registered and that transfers be reported to CDC and conducted under a permitting system. As long as the select agents were not transferred, specific information about the facilities that possessed these agents did not have to be reported (NRC, 2009b).

The determination that the Ames strain of *B. anthracis* was used in the attacks led to a process by which the FBI searched for and acquired samples of known and accessible derivative stocks of that strain for comparison. As noted in Chapter 2, the Ames strain had been widely distributed among laboratories around the world for research and vaccine trials, so the FBI first had to identify all laboratories that maintained stocks of the strain. Next, the Bureau had to obtain samples of these Ames strain derivatives, which would constitute the FBIR and be screened for the presence of the mutant genotypes found in the letters (see Chapter 5). To this end, the FBI prepared and issued a subpoena that included a protocol for the collection and submission of Ames strain samples (Box 6-1). This subpoena was sent in February 2002 to 16 laboratories or facilities in the United States that had been identified as possessing stocks of the Ames strain. (It was subsequently determined that one of these domestic laboratories did not possess the Ames strain[1].) In addition to the subpoenas,

---

[1] According to the FBI (discussion with committee, December 11, 2009), 16 U.S. laboratories were originally identified as candidates for having the Ames strain. According to the DOJ case closing summary, however, only 15 domestic laboratories were confirmed as repositories of the Ames strain (USDOJ, 2010, p. 17). CDC provided a listing of all laboratories registered to work with *B. anthracis*, and FBI investigators created their own list based on CDC select agent transfer records documenting every transfer of anthrax between 1997 and 2001 as well as anthrax inventory records

> **BOX 6-1**
> **Subpoena Protocol for Collection and Submission of**
> **Ames Strain Samples**
>
> **Purpose**: Provide guidance in the preparation and shipment of *Bacillus anthracis* Ames strain to the United States Army Medical Research Institute of Infectious Diseases, Fort Detrick, Maryland. As per FBI request, submit two culture agar slants of each distinct *B. anthracis* Ames strain stock in your possession, which differs in source or in other parameters prescribed by the requesting Agency.
>
> **Materials:**
> Tryptic Soy Agar (TSA) Slants (Remel catalog # 08932, or equivalent)
> Sterile inoculating loops
> Adhesive Labels, permanent, waterproof
> STP-100 Infectious Substance Shipping Container (Saf-T-Pac), or equivalent
>
> **Procedure**:
> 1. Collect each *B. anthracis* Ames strain stock as per your institutional inventory and personal knowledge.
> 2. Prepare a minimum of two TSA slant tubes per stock by prelabeling with permanent waterproof labels. Include the following information on the label: "*B. anthracis* Ames strain." Other designators used by your laboratory, date, and your lab name. Additional information for each stock shall be provided separately.
> 3. A representative sample of each stock shall be used for inoculation of the TSA slants. If the stock is an agar culture, do not use a single colony, but rather use an inoculum taken across multiple colonies. Thawed frozen stocks or other liquid suspensions shall be well mixed prior to transfer of inoculum to the TSA.
> 4. Inoculate each TSA slants in a zig zag manner over the surface of the agar.
> 5. Incubate slants at 35-37C for 12-18 hours to confirm culture growth.
> 6. Individually wrap the slants in packing materials approved for the transport of infectious select agents in accordance with regulations for the shipment of such materials.
>
> *Source:* FBI Documents, Preparing and Shipping TSA Slants for *B. Anthracis* Ames.

consent searches were conducted at USAMRIID in Maryland and at Dugway Proving Ground (DPG) in Utah, and a search warrant was executed at a private company in Ohio (Battelle Memorial Institute) in 2004 to ensure that samples were obtained from every stock derived from the Ames strain in those three facilities. Voluntary submissions were also requested from the three foreign

---

that were subpoenaed from over 100 Biosafety Level 3 (BSL-3) laboratories in the United States. "These records were supplemented by information culled from FBI interviews of scientists working at each of these labs, and through FBI reviews of relevant scientific publications mentioning the Ames strain" (USDOJ, 2010, p. 17).

laboratories (in the United Kingdom, Canada, and Sweden) known to possess the Ames strain, two of which provided samples.

The FBI described their approach to assessing Ames strains internationally in the following way: Complementing this initiative was a separate intensive effort to assess whether a foreign government or a terrorist organization may have gained access to the Ames strain and perpetrated the attacks. An exhaustive initiative, one that continued until the closing of the case, addressed this concern. As with the domestic investigation, it included reviews of published research and transfer records. In addition, it involved an assessment of available intelligence on foreign government and foreign actor capability, meetings with foreign experts, witness interviews, and the collection of Ames strain isolates from certain governments. (FBI/USDOJ, 2011).

Laboratories in possession of stocks derived from the *B. anthracis* Ames strain were directed to obtain samples from each of the stocks in both their institutional and personal inventories for submission to the FBIR. The subpoena asked for little information about the provenance of the samples being submitted, with no specific requirement that such information be provided. For this reason, information about the derivation of the various stocks from the original Ames strain was not standardized and varied greatly in its amount and specificity.

In general, stocks of *B. anthracis* are stored in different ways depending on the physiological state of the bacteria (spores or vegetative cells) as well as the procedures used in a particular laboratory. For example, some microbiological stocks are isolated from single colonies, cultured for only a few additional generations, and then stored in freezers; these stocks are expected to be genetically quite homogeneous. Other stocks exist as confluent lawns of cells on slants or as multiple colonies on agar plates; such stocks have the potential to accumulate genetic diversity if they are kept in these states for long periods. Still other stocks, including the one designated RMR-1029, are mixtures of several independent large-scale culture preparations; such mixtures may also harbor genetic diversity. The sampling protocol in the subpoena directed that a "representative sample" of each stock be provided. It specified different methods for obtaining the sample, depending on how the stock was kept in the laboratory. For agar-based stocks, the submitters were told to take an inoculum from multiple colonies ("do not use a single colony", see Box 6-1), although a precise number was not specified. Samples taken from thawed frozen stocks or liquid suspensions were required to be "well mixed" prior to inoculating the slant that was to be delivered to the FBIR. The submitters were apparently not asked, however, to provide information about which methods they had used to make the inocula for submission nor in what form the stocks were stored. In addition, there was no effort to standardize the number of spores or cells that were submitted. These omissions limit the committee's knowledge (and that of the investigators) about the quality of the samples submitted and, in particular, how well each submission met the requirement of being a "representative sample."

In all, 20 laboratories (15 domestic, two foreign, plus USAMRIID, Dugway, and Battelle) submitted to the FBIR a total of 1,070 (USDOJ, 2010, p. 24) samples of stocks derived from the Ames strain of *Bacillus anthracis*. Of these, 1,059 samples were screened and results reported for the presence or absence of the four mutant genotypes—A1, A3, D, and E (FBI Documents, B2M10D2)—using the assays described in Chapter 5. The FBI told the committee that the other 11 samples were not viable, failed to grow the Ames strain of *B. anthracis*, or failed to grow *B. anthracis* at all. The results and interpretation of this screening are described below.

It is important, however, to recognize not only several inherent limitations of the FBIR collection that make it difficult to assess the evidence in any formal statistical sense, but also the effects of the decision to require growth of the samples before testing for the four mutant genotypes. In addition to issues of representativeness, there are issues with the independence of the samples.

First, statistical analyses typically assume that samples are taken at random from a defined population. The FBI aimed to create a comprehensive repository that encompassed the entire population of stocks derived from the Ames strain, rather than a representative fraction thereof. Given uncertainties in the extent of the entire population of Ames stocks worldwide (highlighted by concern about the possibility of clandestine stocks held by terrorist organizations—see section 3.4.3), the lack of specificity in the subpoena protocol, the uncertainties in compliance with the subpoena protocol, the incomplete information on transfers of Ames-derived stocks between laboratories, and the possibility that some stocks were produced but later destroyed, the repository was unlikely to have been comprehensive.[2] DOJ states in its Summary, "The collection of Ames isolates from laboratories both from the United States and abroad that constitute the FBIR are a comprehensive representation of the Ames strain" (USDOJ, 2010, p. 28). Section 6.5 provides further discussion of this issue, along with the implications of the violation of this assumption of "representativeness" for the statistical inferences

Second, there were complex and varying degrees of genetic relationships among the stocks, reflecting their common descent from the original Ames isolate and the history of transfers, single-colony isolations, and mixtures of materials within and between institutions. As a consequence, some sets of related stocks were likely to be represented by many samples and others by few samples, rendering it impossible to assess the relevant frequencies of genotypes across the population of interest.

Third, statistical analyses are critically dependent on replication to provide measures of the sensitivity and specificity of assays, and such replication should

---

[2] The issue of overseas samples discussion in Chapter 3, section 3.4.3 raises additional questions about the comprehensiveness of the FBIR: but as stated previously, these issues were beyond the scope of this committee.

be part of a protocol at every biological and technological level at which it is feasible. Although scientists were required to submit duplicate samples to the repository, only one sample was analyzed for the genotypes of interest while the other was retained in an archive. Subpoena recipients should have been required to submit three or more samples of each stock, so that replicate genotypic assays could have been performed. Such replication would have allowed a more rigorous assessment of the results of the genotypic assays.

Finally, even for replicate samples from the same stock, other sources of potential variation—such as in the density of cells or the spatial distribution of genotypes in the original stock—might have introduced statistically important dependence across the several assays performed on the samples. Thus, the results obtained from the several genotypic assays cannot be assumed to be independent.

6.3 USE OF THE GENETIC ASSAYS TO TEST FOR THE FOUR GENOTYPES

Either during or after the development of the genetic assays, a decision was made that they should be performed on samples of *B. anthracis* Ames only after the cells in the sample were cultivated once again (in addition to the prior cultivation step by the recipients of the subpoena protocol before submitting the samples). According to the FBI, this further cultivation step was done in part to standardize the nature of the submitted samples prior to the testing (FBI, 2009). In addition, it would have expanded the amount of sample available for DNA preparation and subsequent testing by the different contractor laboratories involved in genetic assay development (FBI, 2009). However, the sensitivity of each of the molecular assays would not have necessarily required this additional cultivation step. Furthermore, the committee notes at least two problems with this additional cultivation step. First, the samples that were submitted to the repository but then failed to grow in culture could not be tested; however, they could have been tested if the genetic assays had been directly applied to the samples. Second, each cultivation step potentially created additional bias in the sample, because not all cell genotypes grow at the same rate under the same conditions (see also Section 5.5.2). Some variant cell types might grow more slowly and fall in abundance relative to the other cell types, such that they are no longer detectable, while others may grow more quickly and become more dominant. The committee was unable to assess the extent to which the benefits of sample cultivation outweighed these potential problems.

6.4 DERIVATION OF RMR-1029 SPORES

The presence of genetic variants in the anthrax letter samples (as described in Chapter 5) and in the spore population in RMR-1029 and in samples known to have come from RMR-1029 (as described below) led the FBI to conclude that "RMR-1029 is the parent material of the evidentiary anthrax spore powder" (USDOJ, 2010, p. 28). Based on their investigation, FBI officials stated that RMR-1029 was produced in 1997 as "a conglomeration of 13 production runs of spores by Dugway, for USAMRIID, and an additional 22 production runs of spore preparations at USAMRIID that were all pooled in this mixture" (FBI, 2008c, p. 55). The resulting 164 liters of spore production were concentrated down to about a liter. Originally, the spore preparation was divided into two one-liter Erlenmeyer flasks, one of which was eventually depleted, so that only one flask remained for subsequent repeated sampling, as is discussed further below (FBI, 2008c, pp. 58-59).

This spore preparation was maintained at USAMRIID, where records indicate that it contained approximately 3×10^{10} spores per ml (USAMRIID, 1997). Between 1998 and 2003, aliquots were withdrawn from RMR-1029 for use in animal studies during the development of anthrax vaccines and therapeutics. The amount of RMR-1029 consumed in the course of these studies is documented in terms of transfers out of the flask (B3D14, B3D16), but given the concentrated nature of the preparation, a very small volume would have provided ample material for use as inocula to produce additional spore preparations such as those that might have been used in the letters.

The production history of the RMR-1029 spore population provides a probable explanation for the presence of subpopulations of genetic variants, including variants that could give rise to visually distinctive colonies (morphotypes). For example, colonies of genetic variants that are defective in spore formation typically have surface textures and coloration that distinguish them from the wild type (see Chapter 5). While it is standard practice in microbiology to prepare new cultures by starting from single-colony purified stocks, this procedure was deemed impractical (Martin, 2010) for the preparation of the large volumes of spores that were combined to produce the RMR-1029 stock. This spore suspension was intended for use as a reference stock, or population, that could be drawn upon for many studies, thereby allowing comparison of results across studies performed at different times. The large number of spores required for an extended set of studies could not, however, be prepared in a single batch (especially given the biosafety level required for work with a dangerous pathogen). Because the fermentors used for growing large batches of spores were not available at USAMRIID, the bulk of the material in RMR-1029 was generated at Dugway.

It is likely that some or all of the genetic variants (including especially those discovered on the basis of atypical colony morphologies) present in RMR-1029 were present in the material provided by Dugway, for the following reasons. First, the DPG material was prepared using inocula that had not been started from a single colony but instead came from stocks that had been obtained from USAMRIID. Bulk material from the stock was then used to inoculate blood-agar plates.

A photograph shown to the committee of a dense lawn of *B. anthracis* grown on such a plate at Dugway reveals the presence of many papillae, small outgrowths of bacteria indicative of mutants that are overgrowing their neighbors in the lawn or have some other distinctive feature (Martin, 2010). The scientist who repeatedly prepared these materials for the multiple production runs told the committee that the presence of numerous papillae was the typical outcome. The committee believes that the presence of these papillae can be taken as evidence that the agar slants already contained mutants, that growth of the bacterial population on the blood agar plates selected for mutants, or both. Because *B. anthracis* cells sporulate on blood agar once they reach high density, the papillae could have been outgrowths of sporulation-defective mutants that continued to grow for several generations after other nonmutant cells stopped growing and formed spores.

Second, according to one DPG scientist, the biological material (lawns, including papillae) scraped from these plates was then used directly to inoculate the fermentors at Dugway (Martin, 2010). This material was collected from the plates after the lawn population had largely converted to spores. Because spores must germinate before growth can resume, unsporulated cells (including from sporulation-defective mutants in the inoculum) would likely have had a growth advantage in the fermentor. Because material prepared at Dugway comprised the bulk of the material that was pooled in the RMR-1029 flask, and because the inocula used to prepare the spores had visible evidence of mutants that may have been defective in spore formation, the committee suggests that at least some of the morphotypes identified in RMR-1029 originated from Dugway. Various biological factors would have affected the resulting presence and abundance of the genetic variants, including their growth rates, germination rates, and sporulation efficiencies under the specific cultivation conditions used as well as the rate at which each variant arises by mutation.

6.5 ANALYSES OF THE REPOSITORY SAMPLES AND STATISTICAL INTERPRETATION OF THE EVIDENCE

The assays developed by the various contractors (Commonwealth Biotechnologies, Inc. [CBI], Midwest Research Institute [MRI], IIT Research Institute [IITRI], and the Institute for Genomic Research [TIGR]) and described in Chapter 5 were used to analyze the samples submitted to the FBIR under the subpoena protocol or gathered through the searches performed by the FBI. The purpose of these assays was to search for the presence of the genetic mutations—A1, A3, D, and E—among the FBIR samples to determine whether any of the samples had a genetic profile that matched that of the evidentiary material and could be a possible source of the material used in the letters.

It is important to reiterate here that the existence in laboratories of the Ames strain of *B. anthracis* dates back only to 1981 and that all samples of the Ames strain at laboratories outside USAMRIID had been derived either directly or indirectly some time between 1981 and 2001 from USAMRIID's stocks of the organism. This fact is of substantial significance and limits a quantitative statistical analysis of the results of the FBIR assays.

The committee reviewed the report provided to the FBI by a contractor who performed statistical analysis of the assay results for the 1,059 viable FBIR samples. The results of the assays were summarized by the contractor in a table (reproduced here as Table 6-1) in the *Statistical Analysis Report*.

The FBI did not seek formal statistical expertise until it had completed the genotype assays of the repository samples. Thus, the utility of the statistical analysis was limited in part by the fact that the experimental design was created without input from statisticians. The FBI contractor analyzed the data for

TABLE 6-1 General Results of the Screening of 1,059 Viable FBIR Samples for the Presence of the Mutation Genotypes, as Summarized by the Statistical Consultant to the FBI

	A1	A3	MRI-D	IITRI-D	E
Positive	35	21	64	58	23
"Neg-u"	—	—	21	—	—
Negative	965	946	940	963	1,008
Inconclusive	20	32	34	24	—
Variant	8	29	—	—	—
Pending	—	—	—	1	—
No *Bacillus* DNA	17	17	—	—	15
No growth	14	14	—	13	13
TOTAL	1,059	1,059	1,059	1,059	1,059

NOTES: MRI-D and IITRI-D refer to the two independent assays for the D mutation genotype developed and performed by the Midwest Research Institute (MRI) and IIT Research Institute (IITRI), respectively.

— indicates that no samples were reported in that category.

"Neg-u" = The meaning of this designation is not clear to the committee. "Neg-u" designation appears in the *Statistical Analysis Report* but it is not defined and does not appear in any of the final reports on D assays submitted to the FBI from MIR and IITRI. Subsequent analysis combined these samples from MRI with MRI's 940 "negative" results.

Variant = sample similar to but not identical to the genotype.

Pending = sample analysis incomplete when *Statistical Analysis Report* submitted to FBI. Subsequent report from MRI indicates that MRI classified the sample as "inconclusive."

Inconclusive (IITRI) = inconsistent results on replicates.

Inconclusive (MRI) = inconsistent results on replicates or no growth or no *Bacillus* DNA. Thus IITRI separated the three categories while MRI classified all three categories as "inconclusive." (The final report from MRI [FBI Documents, B2M8D7] divides 35 "inconclusive" samples into nine "inconclusive analysis" (i.e., "failure to produce a clear reproducible positive indication for the presence of the Morph D deletion"), 19 insufficient DNA, and 7 no growth.

SOURCE: FBI Documents, B2M10D2.

947 of the samples that provided definitive results for all four genotypes. For genotype D, the IITRI data were dropped because they did not provide results in all categories, so analysis of the D genotype was based on the MRI data only. An additional 112 samples were omitted from further investigation because the results with these samples were recorded as "inconclusive," "variant," "no growth," or "no DNA" for one or more of the four genotypes. Tables 6-2 and 6-3 summarize the data from the *Statistical Analysis Report* in two ways, in each case using only the 947 samples that produced definitive results. Table 6-2 shows the frequencies of positive and negative results for each of the four genetic assays, and Table 6-3 shows the frequency of samples that showed a given profile of positive and negative results across the set of assays.

TABLE 6-2 General Results of the Screening of the 947 Samples that Provided Definitive Results for All Four Genotypes

	A1	A3	MRI-D	E
Positive	27	16	51	16
Negative	920	931	896	931
TOTAL	947	947	947	947

SOURCE: FBI Documents, B2M10D2.

TABLE 6-3 Distribution Results for the Four Assays for Genotypes A1, A3, MRI-D, and E in the 947 Samples

A1	A3	MRI-D	E	Count
positive	positive	positive	positive	8
positive	positive	positive	—	2
positive	positive	—	—	2
positive	—	positive	—	4
—	positive	—	positive	3
—	—	positive	positive	2
positive	—	—	—	11
—	positive	—	—	1
—	—	positive	—	35
—	—	—	positive	3
—	—	—	—	876
			TOTAL	947

NOTE: — = negative result
SOURCE: FBI Documents, B2M10D2.

6.5.1 The FBI's *Statistical Analysis Report*

The *Statistical Analysis Report* prepared for the FBI (FBI Documents, B2M10D2) contains five conclusions that are summarized below.

(a) Frequency of 4-positive samples

The fraction of the 947 samples with positive results for all 4 genotypes was 8/947 = 0.00845. This fraction was offered as an estimate for the probability of the occurrence of a 4-positive (++++) sample. ("Retaining only positive and negative results of the four assays, in the 947 repository samples, eight showed simultaneous positive results ++++ for all four assays (i.e., 0.84 percent), with exact 95 percent confidence interval of 0.0037 to 0.0166 (i.e., from 1 in 270 to 1 in 60)" (FBI Documents, B2M10D2).

(b) Dependence among assays

The interdependence among the assay outcomes (co-occurrence of genotypes) was examined using various statistical tests and it was concluded that the occurrence of the four genotypes did not appear to be independent.

Briefly, as presented in the report, if the outcomes were independent, and noting that 27 of the 947 samples showed A1, 16 of the 947 samples showed A3, 51 of the 947 samples showed MRI-D, and 16 of the 947 samples showed E, then one would expect the fraction of samples testing positive for all four genotypes to be 0.4383 per million, or less than one in a million. Since there were 947 samples, one would have expected to see 947 times this fraction, or 0.0004 samples with all four mutations. That is, if the assay outcomes really were independent, one would not expect to have seen *any* samples with all four mutations.[3] In fact, however, the FBIR contained eight samples that were positive for all four mutations, a number considerably larger than the expected 0.0004 samples based on the assumption of independence. Similarly, the report noted that the fraction of samples that would be expected to have only one of any of the four genotypes can be calculated as 0.1072; hence one would expect to see (0.1072 × 947) = 101.5 of the 947 samples with only one positive assay. The FBI observed 50 samples with one positive assay.

(c) Relationships among samples

Given the existence of historical relationships among the 947 repository samples, the *Statistical Analysis Report* described two methods to assess these relationships. The first was a "network analysis" applied to the distribution of combinations of assay results for the four genotypes among the FBIR samples. Table 6-3 provides the number of each of these 11 combinations among the 947

---

[3] Under the assumption that the assay results are independent, the probability of joint outcomes equals the product of the probabilities of each separate outcome. The observed frequencies are estimates of the probabilities. Multiplying the four observed frequencies together yields $0.0285 \times 0.0169 \times 0.05385 \times 0.0169 = 0.4383 \times 10^{-6}$ for 4 positive genotypes, and $0.9715 \times 0.9831 \times 0.94615 \times 0.9831 = 0.8884$ for 4 negative genotypes.

FBIR samples that were considered in the *Statistical Analysis Report*. The network analysis provides a representation of the relationships among the genotypic configurations, informed by their frequencies and the mutational differences among them. In the Report, it was concluded from this representation that the eight samples that were positive for all four genotypes were most closely associated with the two samples that were positive for three genotypes (A1, A3, D) and with the six samples positive for two of these three genotypes. From the analysis, it appeared that the 16 samples positive for E were less closely related.

The second method in the *Statistical Analysis Report* to assess relationships among samples relied on a spreadsheet of dates (rows) and labs (columns) indicating the date when a sample "from" one institution was transferred "to" another facility. (The spreadsheet indicated that transfers occurred but it did not indicated the specific relationships among samples.) The FBI told the committee that the spreadsheet was generated from laboratory reports and included all recorded transfers of Ames samples between 1981 and 2001. The spreadsheet showed known direct or indirect relationships between Laboratory F (USAMRIID) and most (but not all) of the other nine laboratories or institutions that submitted samples that were positive for one or more of the four genotypes. Based on these two methods, the *Statistical Analysis Report* concluded that there were associations between the eight samples testing positive for four genotypes and the other samples that were positive for one or more of the genotypes.

(d) Sensitivity and specificity

The statement of work that preceded the submission of the *Statistical Analysis Report* indicated plans to provide estimates of sensitivities and specificities of the assays for the four genotypes. The committee was able to find only a limited analysis of this sort in the final report.

(e) Significance of "Laboratory F" as the source of seven of the eight samples found to be positive in all four assays

The *Statistical Analysis Report* concluded that the chance of finding eight 4-positive samples should be very low (i.e., less than 0.0166, or less than 1 in 60). The report also notes that seven of the eight 4-positive samples came from only one institution ("Laboratory F"), that the eighth sample came from another institution, and that "its occurrence in [the other laboratory] is explained by a recent sample transfer" (FBI Documents, B2M10D2).

6.5.2 Committee Assessment of *Statistical Analysis Report*

Representativeness, Randomness, Independence

Many of the methods used in the *Statistical Analysis Report* rely on the assumption that the 947 FBIR samples were a representative and random collection, independently sampled from some well-defined population of *B. anthracis* samples. In fact, there is no meaningful population beyond the repository itself, and the repository could not be a random collection of independent samples due to the relationships and sharing among laboratories. Further, the FBIR sometimes contained samples from the same source. The FBI confirmed (1/14/2011 meeting) one such instance: FBIR samples 006-002 (*Statistical Analysis Report*, p.28) and 067-001 (*Statistical Analysis Report*, p. 57) were duplicates made from the same submission ("the disputed sample"). Although the lack of independence among the genotype assays was properly acknowledged in the *Statistical Analysis Report*, statistical calculations were performed that relied on the assumption of independence among the 1,059 samples (e.g., chi-squared tests on p. 10, *Statistical Analysis Report*).

Committee concerns about each part of the *Statistical Analysis Report* are described below:

(a) Frequency of 4-positive samples

Because the eight 4-positive samples and the 947 overall samples were not independent samples, the proportion (8/947) as an estimate for the probability of the occurrence of a 4-positive (++++) sample is not meaningful nor is a calculated "95% confidence interval" for this probability.[4]

(b) Dependence among assays

Table 6-4 from the *Statistical Analysis Report* shows all possible ways of obtaining samples with no positive assays, and 1-, 2-, 3-, or all 4-positive assays. The *Statistical Analysis Report* includes a formal statistical test to assess the significance of the differences between these "observed frequencies" and the "expected frequencies." However, the inference from this test is not valid because, as noted above, the 947 samples were not independent.

The committee identified two potential sources of dependence among repository samples in the results of the four genotype assays. The most obvious is the genealogical structure of descent from the original Ames ancestor in light of the history of transfer of *B. anthracis* samples and stocks among laboratories (see discussion below). Another possible source of dependence among the assay results stems from variation among samples in the process of preparation (under the guidance of the subpoena protocol) and in the amount of DNA, leading to sample-specific variation in probabilities of detection of the genotypes.

---

[4] The confidence interval is computed from the binomial distribution (Snedecor and Cochran, 1989). This calculation is valid only if the 947 samples are independent. As an aside, even if it were appropriate to calculate a 95 percent confidence interval with these data, the calculation is misinterpreted in the Statistical Analysis Report. The calculated interval (0.0037, 0.0166) is actually a 95 percent confidence interval for the probability that a sample from the target population is a 4-positive sample, not for the occurrence of eight 4-positive samples in the target population, as stated in the Statistical Analysis Report (FBI Documents, B2M10D2; cited above in Section 6.1(a)).

TABLE 6-4 Observed and Expected (Under Independence) Distribution of Positive Signatures of Four Genotypes

Number of + signatures	Observed frequency	Expected frequency
0	876	841.2891
1	50	101.4924
2	11	4.1484
3	2	0.0679
4	8	0.0004
TOTAL	947	947.0000

SOURCE: FBI Documents, B2M10D2.

(c) Relationships among samples

Given the counts of the different genotypic configurations in Table 6-3, the result of the network analysis is inevitable and adds no new insights. This frequency-dependent analysis is dominated by the 876 samples showing none of the four genotypes, while the appearance of a more distant relationship of E-positive samples results both from their low frequency and the absence of E in the only 2 samples positive for three genotypes. Note, however, that only 16 of the 23 E-positive samples were included in this analysis, and that several of the 7 other E-positive samples excluded from the statistical analyses were also positive for other genotypes (Appendix C, Table C-1). Further, the network analysis is designed to display relationships among naturally evolving organisms. No conclusions can be drawn from its application to the FBIR samples, without knowing their complex history of transfer and mixing relationships.

(d) Sensitivity and specificity

As noted above, the statement of work that preceded the submission of the *Statistical Analysis Report* included a request for estimations of sensitivity and specificity of assays; however, the committee found only a limited analysis of this sort in the final report. While assay sensitivity and specificity were assessed by the assay development contractors in an idealized and artificial setting during the development of the assays, a more meaningful assessment in the context of the testing of actual repository samples would require replication of tests using FBIR samples, or the creation of multiple types of simulated stock samples.

To assess the sensitivity of the assays, dilution experiments were conducted on two samples, RMR-1029 and "SPS.266 Tube#5," using three replicates at each of 10 dilution levels. Unfortunately, the analysis reveals substantial variation in the assay results. In RMR-1029, the genotypes were usually not detected beyond a dilution of 1:10. For SPS.266 Tube#5, one of the three replicates was positive for E out to a dilution of 1:100,000, yet two replicates

TABLE 6.5 Genotype Assays on Three Replicates from Two Samples at 10 Dilution Levels: Entry Denotes Number of Positive Assays on Three Replicates at Each Dilution Level

Sample 1: RMR-1029

	Dilution level									
	10.1	10.2	10.3	10.4	10.5	10.6	10.7	10.8	10.9	10.10
A1	3	—	—	—	—	—	—	—	—	—
A3	3	1	—	1	—	—	—	—	—	—
D:IITRI	3	—	—	—	—	—	—	—	—	—
D:MRI	3	—	—	—	—	—	—	—	—	—
E	1	1	—	—	—	—	—	—	—	—

Sample 2: SPS.266 Tube#5

	Dilution level									
	10.1	10.2	10.3	10.4	10.5	10.6	10.7	10.8	10.9	10.10
A1	3	1	—	—	—	—	—	—	—	—
A3	3	3	3	2	1	1	—	1	—	—
D:IITRI	3	2	1	—	—	—	—	—	—	—
D:MRI	3	2	1	—	—	—	—	—	—	—
E	1	1	1	1	1	—	—	—	—	—

NOTE: — denotes either negative or no growth
SOURCE: FBI Documents, B2M10D2.

were negative even at just a ten-fold dilution and one replicate for A3 was positive even at a 1:100,000,000 dilution, whereas most were negative at 1:100,000 to 1:10,000,000 dilutions. Table 6-5 provides the results from these dilution experiments summarized in the *Statistical Analysis Report*.

The lack of agreement between MRI and IITRI assays for the D genotype has already been noted (Table 6-1), and further illustrates the differing sensitivities and specificities of the assays. The results of the two assays agreed for 945 of the 1,059 samples, for a concordance rate of 92.1 percent, but disagreed for 30 samples.[5] While concordance is informative, the 30 samples with discordant results provide a valuable opportunity to assess the validity of the assay and gain further insight into these samples; this opportunity was not fully exploited.

---

[5] The decision to set aside the IITRI results seems somewhat arbitrary. The report states that IITRI obtained "no growth" with 13 samples but MRI obtained growth on all samples. However, the Final Report from MRI-D (FBI Documents, B2M8D17), dated 21 July 2006, notes "No growth" on 7 samples. The total number of samples yielding no information was similar for both MRI-D (35) and IITRI-D (37).

In its conclusions, the FBI paid particular attention to samples carrying three or four of the genotypes. However, the FBI did not address the issue of false negative results. In connection with this issue of sensitivity of the assays, a major concern regarding the *Statistical Analysis Report* is the restriction of its analyses to the 947 samples that contained no inconclusive or variant results. Additionally, no errors and uncertainties of detection, nor variation in sample preparation, are taken into account for the analysis. Four of the 112 disregarded samples scored positive for the remaining three genotype assays (see Appendix C, Table C-1).

The lack of replication in the assays of the FBIR samples makes it impossible to quantify the strength of any finding relating to the presence or absence of genotypes in the repository samples since some absences may be false negatives. Because samples were not retested and because the dilution experiments demonstrate the potential for different results on the same sample, one cannot quantify the strength of any finding related to the absence or presence of genotypes in the repository samples: thus, some test results of "negative" could well be false negatives ("present but unable to detect"). Consequently, the finding of all four genotypes in both RMR-1029 and seven samples from one laboratory clearly suggests a relationship between RMR-1029 and three of the four attack materials, but it is impossible to calculate any measure of "statistical strength" for this association.

(e) Significance of Laboratory F as the source of seven of the eight 4-positive samples

As suggested in the *Statistical Analysis Report* (p. 2), on first inspection it may seem noteworthy that seven of the eight samples scoring positive for all four genotypes came from "Laboratory F." However, Laboratory F submitted almost two-thirds (63 percent) of the 947 samples. (Most laboratories submitted fewer than 15 samples; seven submitted 18 to 74 samples, and Laboratory F submitted 598 samples; see Table C-2 in Appendix C.) Consequently, Laboratory F could have submitted most or even all of the eight 4-positive (++++) samples merely by chance. More precisely, the probability that seven, or even all eight, of the 4-positive samples would end up among the Laboratory F 598 submissions, merely by chance, is 0.14223 (about 1 in 7).[6]

6.6 ANALYSES BASED ON RESAMPLING OF RMR-1029 AND INTERPRETATION OF THE RESULTS

In late February 2002, Bruce Ivins and a technician who worked with him at USAMRIID prepared duplicate samples of four Ames-derived stocks from

---

[6] This probability calculation arises from the hypergeometric distribution for the probability that 7 or 8 of the 8 samples arise in a subset of 598 samples from the total 947 samples; see Appendix C, Table C-2.

his laboratory for submission to the FBIR. The second copy of each sample was sent to Paul Keim's laboratory at Northern Arizona University. According to the Department of Justice (USDOJ, 2010, p. 78), on or before March 28, 2002, the USAMRIID staff who had collected that laboratory's Ames strain samples for submission to the FBI "advised Dr. Ivins and his laboratory technician that their submissions were not prepared according to the FBIR protocol." The DOJ report continues: "Specifically, Dr. Ivins and his lab technician used homemade slants as opposed to the commercially available Remel slants specified by the protocol, so the four slants prepared on February 27, 2002 were rejected by the FBIR, and Dr. Ivins was told to resubmit his culture samples on the appropriate slants" (p. 78). The FBI disposed of its rejected samples but, importantly, the duplicate copies sent to Keim's laboratory were retained, including one that had come from flask RMR-1029 and that would later be shown to contain all four genotypes. In April 2002, Ivins submitted four newly prepared samples, again in duplicate, and this time the samples were accepted. One copy of each sample entered the FBIR and the duplicates were again sent to Northern Arizona University.

In the course of the investigation, certain doubts were raised about whether Ivins had submitted samples of all relevant Ames stocks in his possession to the FBIR. In April 2004, the FBI secured RMR-1029 and other *B. anthracis* samples that had been under Ivins's control, and these were transferred elsewhere for various tests. From these tests, the FBI concluded that: "Genetic analysis determined that . . . RMR-1029—the purest and most concentrated batch of Ames spores known to exist—was the parent to the evidentiary material used in the anthrax mailings" (USDOJ, 2010, p. 79). The committee addressed this conclusion (see discussion above) and also focused on secondary analyses performed in order to reconcile certain discrepancies in the multiple samples that were submitted by Ivins and purported to come from flask RMR-1029.

FBI investigators had observed that the second RMR-1029 sample submitted by Ivins in April 2002 did not score positive for any of the four genotypes discovered in the attack letters and in RMR-1029 as well as in samples derived from RMR-1029 that were submitted by other scientists. Based on this apparent discrepancy, in late 2006 the FBI obtained from Northern Arizona University the duplicate of the first RMR-1029 sample that Ivins had submitted in February 2002, which was then put into the FBIR and analyzed. This earlier RMR-1029 sample scored positive for all four of the genotypes that were assayed (A1, A3, D, and E), whereas the later sample had scored positive for none of them.

The FBI sought to determine the cause of this discrepancy between the earlier and later submissions by Ivins that both were supposed to have come from the same RMR-1029 flask. One possibility (the null hypothesis) was that repeated samples from RMR-1029 following the FBI protocol might produce variable results. This possibility could reflect, for example, insensitivity of the molecular tests at the genotypic frequencies present in that flask. To address that issue, and on the recommendation of external science advisors (the "Red Team"; USDOJ, 2010, p. 79), the FBI directed that an experiment be performed in which the RMR-1029 flask was sampled in an identical manner 30 times following the subpoena instructions. These replicate samples were processed at the National Bioforensic Analysis Center and the resulting material analyzed by scientists at CBI, MRI, IITRI, and the University of Maryland for the presence of the same four genotypes. According to the DOJ summary of the case (USDOJ, 2010, p. 79), the results were as follows: "Occasionally, only three of the four genetic mutations were detected, and at no time were less than three detected. It followed that if Dr. Ivins prepared his submission to the repository in accordance with the protocol, that submission could not miss all four of the morphological variants present in RMR-1029."[7]

Table 6-6 presents the results of these 30 analyses. These data demonstrate substantial variability among the 30 replicates. Sixteen of the 30 samples scored positive for all four mutations and another eight scored positive for three of the four mutations; however, five samples scored positive for only two mutations, and one sample (Sample 20) scored positive for just one of the four mutations. (Inconclusive results, as well as cases where the IITRI and MRI tests for the D genotype gave different results, were considered "negative" outcomes by the committee.) None of the 30 samples scored negative for all four mutations.

Given these assay results of the 30 replicates, what is the probability that analysis on an additional sample, taken from RMR-1029 in the same manner, would yield negative results for all four genotypes by chance alone? Since none of the 30 samples was negative for all four assays, the probability of obtaining a sample of four negatives, by chance alone, cannot be very high and, according to the binomial distribution[8] it is unlikely to be higher than 0.095 (9.5 percent).

At the January 2011 meeting with the FBI, the committee was told that the second copy of this disputed submission was also analyzed and tested negative for all four genetic markers. That is, the second copy of Ivins' disputed April 2002 submission, kept at NAU along with the second copies of all submissions, was moved into the FBIR, given a new identification number, retested, and found to be negative for all four genetic markers. If the assays on these two samples were as independent of one another as the assays performed on the 30 samples in Table 6-6, then the probability that both sets of assays would yield

---

[7] On October 15, 2010, The DOJ issued an erratum stating: "Based on a 'Red Team' recommendation, experiments were prepared at the direction of the FBI Lab to address the FBIR submission process with regard to RMR-1029. RMR-1029 was sampled 30 times in accordance with the subpoena instructions. In a few instances, fewer than three markers were detected. However, in none of the 30 attempts were no markers detected. It followed that if Dr. Ivins prepared his submission to the repository in accordance with the protocol, that submission could not miss all four of the morphological variants present in RMR-1029."

[8] That is, if the probability of a sample having four negatives were greater than 0.095, the probability of obtaining no such samples among the 30 would be less than $(1 - 0.095)^{30} = 5\%$.

COMPARISON OF THE MATERIAL WITH SAMPLES IN THE FBI REPOSITORY 143

TABLE 6-6 Results Obtained by Resampling from Flask RMR-1029

	A1	A3	D IITRI	D MRI	E
Sample 1	+	+	+	+	+
Sample 2	+	+	+	+	+
Sample 3	+	+	+	+	+
Sample 4	+	+	+	+	+
Sample 5	+	+	+	+	+
Sample 6	+	+	+	+	+
Sample 7	+	+	Negative	+	+
Sample 8	+	+	+	+	+
Sample 9	+	+	+	+	+
Sample 10	Inconclusive	+	Inconclusive	Inconclusive	+
Sample 11	+	+	+	+	+
Sample 12	Inconclusive	+	Inconclusive	Negative	+
Sample 13	Inconclusive	+	Negative	Inconclusive	+
Sample 14	Negative	+	+	+	+
Sample 15	Negative	+	+	+	+
Sample 16	+	+	+	+	+
Sample 17	+	+	+	+	+
Sample 18	Negative	+	+	+	+
Sample 19	+	+	+	+	+
Sample 20	Negative	Inconclusive	Negative	Negative	+
Sample 21	+	+	+	+	+
Sample 22	Negative	+	+	+	+
Sample 23	+	+	+	+	+
Sample 24	Negative	+	Negative	Negative	+
Sample 25	Inconclusive	+	+	+	+
Sample 26	+	+	+	+	+
Sample 27	Negative	+	+	+	+
Sample 28	+	+	+	+	+
Sample 29	Negative	+	+	+	+
Sample 30	Negative	+	Inconclusive	Negative	+

NOTES: += positive ; IITRI = IIT Research Institute; MRI = Midwest Research Institute
SOURCE: FBI Documents, B2M10D2.

negative results for all four genotypes would be (0.095 × 0.095) = 0.009025 (or 0.9 percent). While it is still possible that two such results could have occurred by chance alone, this chance is very small (less than one percent).

6.7 COMMITTEE FINDINGS

During the course of the committee's review of the scientific evidence, the Department of Justice officially closed the investigation. The committee is not in a position to offer a judgment about the importance and strength of the scientific investigation relative to the importance and strength of the criminal investigative component of this case because it was not charged with (and lacked the expertise for) reviewing the latter. Our major finding is that:

It is not possible to reach a definitive conclusion about the origins of the *B. anthracis* in the mailings based on the available scientific evidence alone.

Finding 6.1: The FBI appropriately decided to establish a repository of samples of the Ames strain of *B. anthracis* then held in various laboratories around the world. The repository samples would be compared with the material found in the letters to determine whether they might be the source of the letter materials. However, for a variety of reasons, the repository was not optimal. For example, the instructions provided in the subpoena issued to laboratories for preparing samples (i.e., the "subpoena protocol") were not precise enough to ensure that the laboratories would follow a consistent procedure for producing samples that would be most suitable for later comparisons. Such problems with the repository required additional investigation and limit the strength of the conclusions that can be drawn from comparisons of these samples and the letter material.

The FBI and contract scientists appropriately recognized that the mutations in the letter isolates provided information that might help identify the source of the *B. anthracis* used in the attacks, developed appropriate assays for four of these mutations, and created and screened a repository of Ames strain samples. Based on the results of that screening, FBI scientists appropriately concluded that the majority of repository samples contained none of the four mutations, although 50 of the samples contained one of the four mutations and 10 samples had three or all four mutations (the numbers with one or more mutations are higher if one includes samples that were excluded in the FBI's statistical report). However, features of the repository including unknown sample provenance, and the history of sharing and mixing of stocks, presented investigative challenges.

The first challenge with the repository was the lack of independence among samples and an incomplete understanding of the provenance of samples due to the known history of sharing. Sharing of samples between laboratories is an important part of scientific research and is critical to testing reproducibility and furthering scientific analysis. Prior to the attacks of 2001, several institutions shared samples of the Ames strain to different extents, resulting in variation in the numbers and kinds of samples they submitted to the FBIR. Also, this sharing extended not only to substrains but also to mixtures of several substrains that were grown in separate batches and then pooled. While such sharing and transfers are important in scientific research, they make it more difficult to identify a unique source of the mutations found in the attack materials. In recognition of this important issue, FBI scientists and investigators sought to determine the history of shipments among institutions and the genealogical relationships among samples in the repository, but they never obtained a complete record.

Another challenge with the repository was that, since the importance of the mutant genotypes was not fully understood when the subpoena protocol was written, the document was vague (e.g., "use an inoculum taken across multiple colonies"), and was not written in a way that would maximize the chance that variant genotypes in a mixed stock population would be submitted. Thus, if the four assayed genotypes had been present in a laboratory culture at low frequency, it is not clear whether they would have found their way into the sample of the culture submitted to the repository, since as few as two colonies would have satisfied the instructions provided in the subpoena protocol. After the importance of the mutant genotypes became known, there was no request for additional samples using a revised protocol that might have improved the sampling.

A final challenge was that the repository collection process was based on the integrity of the individuals asked to provide samples. If the motive for the repository was to identify the source of the letter material, standards of custody of evidence would dictate that agents of the FBI should have obtained the samples. In most instances, holders of the material were asked to provide samples and send them in. The sender could have been the instigator and may not have complied with instructions, as the FBI alleges with respect to Dr. Ivins.

Finding 6.2: The results of the genetic analyses of the repository samples were consistent with the finding that the spores in the attack letters were derived from RMR-1029, but the analyses did not definitively demonstrate such a relationship.

The scientific data alone do not support the strength of the government's repeated assertions that "RMR-1029 was conclusively identified as the parent material to the anthrax powder used in the mailings" (USDOJ, 2010, p. 20), nor statements about the role of the scientific data in arriving at their conclusions, as in "the scientific analysis coordinated by the FBI Laboratory determined that RMR-1029, a spore-batch created and maintained at USAMRIID by Dr. Ivins, was the parent material for the anthrax used in the mailings" (USDOJ, 2010, p. 8).

The committee agrees that the genetic evidence is consistent with and supports an association between the RMR-1029 flask and the *B. anthracis* used in the three attack letters that were tested; however, there are several important caveats. As discussed above, the nature of the repository collection, including the incompletely documented history of sharing and mixing of Ames strain stocks and the ambiguity in the subpoena protocol, makes it difficult to quantify the strength of the evidence linking RMR-1029 to the letters, because of the complex and ill-defined nature of the reference population. The materials from at least three of the four attack letters harbored a mixture of several variant genotypes. By contrast, the repository may include samples recently derived from single colonies, as is customary for microbiological research, but it is unlikely that these samples by virtue of this recent history would have more than one of the variant types. Moreover, owing to the practice of sharing samples among laboratories, many samples were essentially duplicates of one another. It is also possible that the sample repository was incomplete because the global distribution of Ames stocks was not known or because some stocks might have been destroyed prior to the subpoena. These limitations made it impossible for the committee to generate any meaningful estimate of the probability of a coincidental match between the *B. anthracis* genotypes discovered in the attack letters and those later found by screening samples from the RMR-1029 flask.

Finding 6.3: Some of the mutations identified in the spores of the attack letters and detected in RMR-1029 might have arisen by parallel evolution rather than by derivation from RMR-1029. This possible explanation of genetic similarity between spores in the letters and in RMR-1029 was not rigorously explored during the course of the investigation, further complicating the interpretation of the apparent association between the *B. anthracis* genotypes discovered in the attack letters and those found in RMR-1029.

Another challenge with determining the cause of the apparent association between some of the *B. anthracis* genotypes in the attack letters and those found in RMR-1029 stems from the possibility that the same mutations might have arisen repeatedly in other Ames strain populations. Colony variants with similar or even identical mutations might arise repeatedly in other populations for two reasons. First, we do not know the number and rate of possible mutations that could produce similar phenotypes. Research by Worsham and colleagues (see, for example, Worsham and Sowers, 1999) identified numerous oligosporogenous variants with phenotypes similar to those described in the letters. In response to questions raised by the committee, the FBI indicated that among the 296 Ames submissions to the FBIR by Worsham, only the Morph D genotype was detected, and it was in three samples. The A1, A3 and E mutations were not detected in any of the Worsham samples submitted to the FBIR (FBI, 2010a). Second, under the conditions used to grow *B. anthracis* for the large scale production of spores, there may well have been inadvertent selection that favored oligosporogenic mutants, which would cause the frequency of these mutants to be higher than expected for mutations that conferred no advantage to the cells, thereby increasing the likelihood of parallel evolution in replicate spore productions. The recent published work by Sastella and colleagues (2010) highlights the possible role of repeated passage of *B. anthracis* in the enrichment of sporulation-deficient mutants.

Finding 6.4: The genetic evidence that a disputed sample submitted by the suspect came from a source other than RMR-1029 was weaker than stated in the Department of Justice, *Amerithrax Investigative Summary.*

The committee reexamined the data that the FBI obtained following the discovery that one of the samples submitted by Bruce Ivins, which was supposed to have been taken from RMR-1029, did not test positive for any of the four assayed mutations (A1, A3, D, and E) in either of two copies analyzed. As discussed in Section 6.6, an experiment was performed in which 30 replicate samples were taken from RMR-1029 according to the FBI subpoena protocol and tested for the four mutations. Based on these results, the committee found that it is, in fact, possible that the disputed sample came from RMR-1029, and the probability of this outcome—that an actual sample from RMR-1029 would test negative for all four genotypes in two sets of assays—might be on the order of 1 percent. Hence, while the evidence is strongly suggestive that the disputed sample was not taken from RMR-1029, it is less certain than is indicated in the original version of the case-closing summary issued by the DOJ, which asserted that all 30 additional samples scored positive for at least three of the four genotypes, and concluded that "It followed that if Dr. Ivins prepared his submission to the repository in accordance with the protocol, that submission could not miss all four of the morphological variants present in RMR-1029" (USDOJ, 2010, p. 79).[9]

Finding 6.5: The scientific data generated by and on behalf of the FBI provided leads as to a possible source of the anthrax spores found in the attack letters, but these data alone did not rule out other sources.

The committee was not charged with reviewing, nor was it given access to, the findings from the criminal investigation component of this case. The committee therefore could not assess the potential value of additional scientific investigation with respect to better establishing the source of the *B. anthracis*

---

[9] See footnote 7.

attack spores. Additional experiments might be of value were there a need to strengthen the scientific aspects of the case (in particular, if the case depended on the findings of the scientific investigation component as opposed to the criminal evidence).

Finding 6.6: Point mutations should have been used in the screening of evidentiary samples.

The FBI chose to study only a subset of the mutational variants of *B. anthracis* found in the attack letters. In particular, certain point mutations were not investigated, apparently because FBI scientists regarded them as less stable, more difficult to screen for their presence in the repository, or both. In the committee's view, key evidentiary samples in the repository (along with appropriate controls) should have been screened for all of the mutations found in all of the letters. The stability of point mutations should not have been a concern, as most point mutations have extremely low rates of reversion and most of the methods and data used to track the spread of infections rest on this stability. Moreover, the forward rate of deletion mutations and both the forward and reverse rates of insertion mutations (the types of mutations screened for in the A1, A3, D, and E genotypic assays; see Chapter 5 and Table 5.2) are usually higher than the corresponding rates for point mutations. Thus, the possibility that identical mutations may arise independently (in parallel) is greater for insertions and deletions than for point mutations. With regard to the concern about the greater difficulty of screening for point mutations, it should be feasible to sequence directly the relevant genes in a large number of samples, including those that are genotypic mixtures such as RMR-1029 and the letter samples, using high-throughput "next-generation sequencing" methods (see Finding 6.8).

Finding 6.7: Biological material from all four letters should have been examined to determine whether they each contained all four genetic markers used in screening the repository samples.

The FBI observed morphological variants in the *B. anthracis* isolated from the attack letters and used those variants to identify mutations that were then used as genetic markers during the systematic screening of the repository samples. However, the FBI did not systematically examine the materials from the attack letters to determine whether all of these markers were present in each letter sample (possibly owing to concerns about the limited amount of biological material from some of the letters). And although genotypes A1, A3, and E were identified in all three letters that were examined (*New York Post*, Leahy, Daschle), the D genotype was found only in the *New York Post* letter. The material in the Brokaw letter was not examined for the presence of any of the genotypes, so the FBI could only infer from other nongenetic evidence that the biological properties of the materials in this letter were the same as or similar to the materials in the *New York Post*, Leahy, and Daschle letters.

In the committee's view, it would have been useful to determine whether all of the genotypes were present in all of the letters and, if so, at what relative abundances. This issue is important because similar morphotypes can and did arise from different mutations, and multiple mutations producing similar phenotypes were present in some of the letters. Thus, the presence of similar morphotypes in different letters does not mean that the same genotypes were present in all the letter samples. Given the conspicuous differences in the physical properties of some of the letter samples, it is even more important to establish their genetic similarities and, if relevant, consider the possible implications of any genetic differences that might have been found.

Finding 6.8: New scientific tools, methods, and insight relevant to this investigation became available during its later years. An important example is high-throughput "next-generation" DNA sequencing. The application of these tools, methods, and insight might clarify (strengthen or weaken) the inference of an association between RMR-1029 and the spores in the attack letters. Such approaches will be important for use in future cases.

When the committee began its deliberations the FBI's anthrax letters investigation was still open, although the FBI had publicly declared its confidence in having identified the sole person responsible for the mailings. During the committee's review of the scientific evidence, the DOJ officially closed the investigation. The committee is not in a position to offer a judgment about the importance and strength of the scientific investigation evidence relative to the importance and strength of the criminal investigation evidence, because it was not charged with (and lacked the expertise for) reviewing the latter.

Since 2001, important technological advances have occurred that would allow a more thorough and systematic analysis of genetic similarities and differences between key evidentiary samples, repository samples, and appropriate controls. When the investigation began, bacterial genomes were typically sequenced at an average coverage of fewer than 10 reads per nucleotide (see, for example, Read et al., 2002). Given the error rates inherent in sequencing technologies and the costs of whole-genome sequencing at that time, it was feasible to sequence only a few well-chosen clones as was done first to compare an isolate from the Florida victim with the Porton Down Ames strain (Read et al., 2002), and then search for mutations in several morphotypes (as described in Chapter 5).

It was also not easy in the early 2000s to sequence and interpret the data from mixtures of bacterial genomes. However, that changed during the course of the investigation and it is now possible to sequence bacterial genomes, even including minority components of heterogeneous populations, to much greater depths of coverage much more quickly, allowing discovery of genetic polymorphisms (mutational variants) in bacterial populations without requiring phenotypic discrimination (Barrick and Lenski, 2009; Holt et al., 2009). Appropriate statistical methods must then be used to distinguish variations that reflect errors in sequencing from mutations actually present in the mixture. Importantly, any application of next-generation sequencing technologies to samples and evidentiary material from this case would have required that the methods be validated for use with samples and material of this type and for the intended purposes and questions at hand.

With such technologies and methods, one would likely have discovered additional polymorphisms in RMR-1029 and the letter samples. To the extent that any new polymorphisms were found to be concordant between these evidentiary materials (within statistical limits), that would strengthen the genetic association between the bacteria from the letters and those from the suspect flask. It is also conceivable, however, that such additional analyses might have revealed further mutations in the samples from the attack letters that were not present in RMR-1029, thereby weakening the link between the evidentiary material and RMR-1029. Thus, while it was not feasible at the start of the investigation, investigators should have subsequently examined the value from a forensic perspective of "deep sequencing" of key samples. Given the limitations of the existing repository, it is uncertain whether this further scientific investigation would have identified an altogether different source of the *B. anthracis* attack spores, but it could have provided additional information on the process used to generate the material in the attack letters.

More generally and looking forward, the committee anticipates that deep-sequencing methods (including, as appropriate, metagenomic analyses of environmental samples of diverse microbial communities) will be an important forensic tool in future investigations of any similar event involving microbial pathogens.

It should be noted that future biological attacks may pose even greater challenges than did this attack. For example, the biological agent may belong to a species with a more complex and less well understood population structure, it may be genetically modified in a manner that obscures its origin, or a direct sample of the attack material may not be available. This last possibility means that environmental and clinical samples, with their additional challenges, may have greater importance in a future investigation.

Finding 6.9: The FBI faced a difficult challenge in assembling and annotating the repository of *B. anthracis* Ames samples collected for genetic analysis.

Much of the challenge in assembling and annotating the repository was inherent to the types of materials involved, which included stocks that had been derived, sampled (or combined), and then stored for different periods in several ways and for diverse scientific purposes in many laboratories. The FBI collected substantial information, or metadata, about many or all of these samples including details of historical derivation, mode of sampling and storage, and so forth. As was appropriate, the genetic screening of the samples in the collection was performed in a "blind" fashion so that this information would not influence the test results. If a future bioterrorism event requires the establishment of a sample repository, attention should be paid to the structure of the database and the inclusion of (or ways to link to) any and all relevant metadata.

Finding 6.10: The evidentiary material from this case is, and will be, immensely valuable, especially in the event of future work on either this case or other cases involving biological terrorism or warfare. It is critically important to continue to preserve all remaining evidentiary material and samples collected during the course of this (the anthrax letters investigation) and future investigations, including the overseas environmental samples, for possible additional studies.

Recent and future advances in scientific methods and insight may provide the means to extract additional valuable information from case-associated material and samples. In addition, in the event of a future biological attack, these materials and samples may prove useful for comparative analyses. Therefore, despite the closure of this case, all remaining case-associated materials and samples should be retained and preserved for possible further studies.

Bibliography

Alberts, B., Johnson, A., Lewis, J., Raff, M., Roberts, K., and P. Walter. 2002. *Molecular Biology of the Cell,* 4th ed. New York: Garland Science.
Andersen, G.L., Simchock, J.M., and K.H. Wilson. 1996. Identification of a region of genetic variability among *Bacillus anthracis* strains and related species. *J Bacteriol,* 178(2):377-384.
Atlas, R.M. 2002. Responding to the threat of bioterrorism: A microbial ecology perspective—the case of anthrax. *Int Microbiol,* 5(4):161-167.
Baron, P.A., Estill, C.F., Deye, G.J., Hein, M.J., Beard, J.K., Larsen, L.D., et al. 2008. Development of an aerosol system for uniformly depositing *Bacillus anthracis* spore particles on surfaces. *Aerosol Sci Technol,* 42(3):159-172.
Barrick, J.E., and R.E. Lenski. 2009. Genome-wide mutational diversity in an evolving population of *Escherichia coli. Cold Spring Harb Symp Quant Biol,* 74:119-129.
Beecher, D. 2006. Forensic application of microbiological culture analysis to identify mail intentionally contaminated with *Bacillus anthracis* spores. *Appl Environ Microbiol,* 72:5304-5310.
Beyer, W., Glöckner, P., Otto, J., and R. Böhm. 1995. A nested PCR method for the detection of *Bacillus anthracis* in environmental samples collected from former tannery sites. *Microbiol Res,* 150(2):179-186.
BioOne. 2005. Country's Final Anthrax Decontamination to be Completed This Month by BioOne. Press Release. Retrieved June 17, 2010, from http://www.bioone.com/2005_0323.htm.
Bouzianas, D.G. 2009. Medical countermeasures to protect humans from anthrax bioterrorism. *Trends Microbiol,* 17(11):522-528.
Bozue, J., Moody, K.L., Cote, C.K., Stiles, B.G., Friedlander, A.M., Welkos, S.L., et al. 2007. *Bacillus anthracis* spores of the bclA mutant exhibit increased adherence to epithelial cells, fibroblasts, and endothelial cells but not to macrophages. *Infect Immun,* 75(9):4498-4505.
Brahmbhatt, T.N., Janes, B.K., Stibitz, E.S., Darnell, S.C., Sanz, P., Rasmussen, S.B., et al. 2007. *Bacillus anthracis* exosporium protein BclA affects spore germination, interaction with extracellular matrix proteins, and hydrophobicity. *Infect Immun,* 75(11):5233-5239.
Brookmeyer, R., Blades, N., et al. 2001. The statistical analysis of truncated data: Application to the Sverdlovsk anthrax outbreak. *Biostatistics,* 2(2):233-247.
Brookmeyer, R., Johnson, E., et al. 2005. Modelling the incubation period of anthrax. *Stat Med,* 24(4):531-542.
Budowle, B. 2009. University of North Texas Health Science Center. Presentation to the committee, July 31.
Budowle, B., Schutzer, S.E., Burans, J.P., Beecher, D.J., Cebula, T.A., Chakraborty, R., et al. 2006. Quality sample collection, handling, and preservation for an effective microbial forensics program. *Appl Environ Microbiol,* 72(10):6431-6438.

Bush, L.M., Abrams, B.H., Beall, A., and C.C. Johnson. 2001. Index case of fatal inhalational anthrax due to bioterrorism in the United States. *N Engl J Med,* 345(22):1607-1610.

Carrera, M., Zandomeni, R.O., Fitzgibbon, J., and J.L. Sagripanti. 2007. Difference between the spore sizes of *Bacillus anthracis* and other *Bacillus* species. *J Appl Microbiol,* 102(2):303-312.

CBS News.com. 2008. Anthrax Suspect Reportedly Commits Suicide: Top Biodefense Researcher Knew Justice Department Was About to File Charges. Retrieved June 17, 2010, from http://wbztv.com/national/anthrax.scientist.suicide.2.785279.html.

CDC (Centers for Disease Control and Prevention). 2001a. Ongoing Investigation of Anthrax—Florida, October 2001. *MMWR Morb Mortal Wkly Rep,* 50(40):877.

CDC. 2001b. Update: Investigation of Anthrax Associated with Intentional Exposure and Interim Public Health Guidelines, October 2001. *MMWR Morb Mortal Wkly Rep,* 50(41):889-893.

CDC. 2001c. Update: Investigation of Bioterrorism-Related Anthrax and Interim Guidelines for Exposure Management and Antimicrobial Therapy, October 2001. *MMWR Morb Mortal Wkly Rep,* 50(42):909-919.

Cole, L.A. 2009. *The Anthrax Letters: A Bioterrorism Expert Investigates the Attacks that Shocked America*. New York: Skyhorse Publishing.

Colwell, R. 2009. University of Maryland College Park and Johns Hopkins University Bloomberg School of Public Health. Presentation to the committee, September 24.

Cybulski, R.J., Jr., Sanz, P., Alem, F., Stibitz, S., Bull, R.L., and A.D. O'Brien. 2009. Four superoxide dismutases contribute to *Bacillus anthracis* virulence and provide spores with redundant protection from oxidative stress. *Infect Immun,* 77(1):274-285.

Didenko, V.V. 2001. DNA probes using fluorescence resonance energy transfer (FRET): Designs and applications. *Biotechniques,* 31(5):1106-1116, 1118, 1120-1121.

DPG (United States Army Dugway Proving Ground). 2004. SEM Photos of *B. antracis* [sic] Ames Production, Dugway Proving Grounds [sic]. (B1M13D4).

DPG. 2006. Final Report for the Analytical Chemistry Analysis of Anthrax Powders, DTC Project Number 8-CO-480-000-0068. February 1. (B1M13D3).

Driks, A. 2009. The *Bacillus anthracis* spore. *Mol Aspects Med,* 30(6):368-373.

Dull, P.M., Wilson, K.E., et al. 2002. *Bacillus anthracis* aerosolization associated with a contaminated mail sorting machine. *Emerg Infect Dis,* 8(10):1044-1047.

Easterday, W.R., Van Ert, M.N., Simonson, T.S., Wagner, D.M., Kenefic, L.J., Allender, C.J., et al. 2005a. Use of single nucleotide polymorphisms in the *plcR* gene for specific identification of *Bacillus anthracis*. *J Clin Microbiol,* 43(4):1995-1997.

Easterday, W.R., Van Ert, M.N., Zanecki, S., and P. Keim. 2005b. Specific detection of *Bacillus anthracis* using a TaqMan mismatch amplification mutation assay. *Biotechniques,* 38(5):731-735.

Elena, S.F., and R.E. Lenski. 2003. Evolution experiments with microorganisms: The dynamics and genetic bases of adaptation. *Nat Rev Genet,* 4(6):457-469.

Ember, L.R. 2006. Anthrax sleuthing: Science aids a nettlesome FBI criminal probe. *Chem Eng News,* 18(49):47-54.

FBI (Federal Bureau of Investigation). 2011. Communication to the committee, January 3.

FBI. 2010a. Communication to the committee, January 13.

FBI. 2010b. Communication to the committee, January 25.

FBI. 2009. Presentation to the committee, September 24.

FBI. 2008a. The Search for Anthrax, Retrieved September 12, 2010, from http://www.fbi.gov/about-us/history/famous-cases/anthrax-amerithrax/the-search-for-anthrax.

FBI. 2008b. Anthrax Investigation: Closing a Chapter, August 6. Headline Archives. Retrieved June 16, 2010, from http://www.fbi.gov/page2/august08/amerithrax080608a.html.

FBI. 2008c. Press Conference for Scientific Media with Dr. Vahid Majidi, Assistant Director of the FBI Weapons of Mass Destruction Directorate and Dr. D. Christian Hassell, FBI Laboratory Director Regarding the Science of the Anthrax Investigation. FBI National Press Office, August 18. Washington, D.C.

FBI/USDOJ (United States Department of Justice). 2011. Presentation to the committee, January 14.

Fenselau, C.C. 2005. Forensic Science and Counterterrorism. Presentation before 17th Sanibel Conference on Mass Spectrometry.
Fisher, B.A.J. 2005. *Techniques of Crime Scene Investigation*, 7th ed. Boca Raton: CRC Press
Fitch, J.P., Raber, E., Imbro, D.R. 2003. Technology challenges in responding to biological or chemical attacks in the civilian sector. *Science*, 302:1350-1354.
FoxNews.com. 2001. $1 Million Reward Offered for Information on Anthrax Terrorists. Retrieved June 17, 2010, from http://www.foxnews.com/story/0,2933,36761,00.html.
Fraser-Liggett, C. 2009. Institute of Genome Sciences and University of Maryland School of Medicine. Presentation to the committee, July 31.
Freed, D. 2010. The Wrong Man. *Atlantic Magazine*. May.
Freedman, D., Pisani, R., and R. Purves. 2007. *Statistics,* 4th ed. New York: W.W. Norton.
Friedlander, A.M., and S.F. Little. 2009. Advances in the development of next-generation anthrax vaccines. *Vaccine, 27 Suppl 4*, D28-D32.
Gallucci-White, G. 2008. Detrick Anthrax Scientist Commits Suicide as FBI Closes In. *Frederick News-Post*, August 1.
Gardner, R.M. 2005. *Practical Crime Scene Processing and Investigation*. Boca Raton, FL: CRC Press.
Greene, C.M., Reefhuis, J., Tan, C., Fiore, A.E., Goldstein, S., Beach, M.J., et al. **2002.** Epidemiologic investigations of bioterrorism-related anthrax, New Jersey, 2001. *Emerg Infect Dis,* 8(10):1048-1055.
Guillemin, J. 1999. *Anthrax: The Investigation of a Deadly Outbreak*. Berkeley: University of California Press.
Gwertzman, B. 1980. *The New York Times*, March 19, p. 1.
Hassell, C. 2009. Federal Bureau of Investigation. Presentation to the committee, July 30.
Heine, H. 2010. Formerly United States Army Medical Research Institute for Infectious Diseases. Presentation to the committee, April 22.
Henderson, I., Yu, D., and P.C. Turnbull. 1995. Differentiation of *Bacillus anthracis* and other "*Bacillus cereus* group" bacteria using IS231-derived sequences. *FEMS Microbiol Lett,* 128(2):113-118.
Henriques, A.O., and C.P. Moran, Jr. 2007. Structure, assembly, and function of the spore surface layers. *Annu Rev Microbiol,* 61:555-588.
Hirota, R., Hata, Y., Ikeda, T., Ishida, T., and A. Kuroda. 2010. The silicon layer supports acid resistance of *Bacillus cereus* spores. *J Bacteriol,* 192(1):111-116.
Hoch, J.A. 2000. Two-component and phosphorelay signal transduction. *Curr Opin Microbiol,* 3(2):165-170.
Hoffmaster, A.R., Fitzgerald, C.C., Ribot, E., Mayer, L.W., and T. Popovic. 2002. Molecular subtyping of *Bacillus anthracis* and the 2001 bioterrorism-associated anthrax outbreak, United States. *Emerg Infect Dis,* 8(10):1111-1116.
Hogan, W.R., Cooper, G.F., et al. 2007. The Bayesian aerosol release detector: An algorithm for detecting and characterizing outbreaks caused by an atmospheric release of *Bacillus anthracis*. *Stat Med,* 26(29):5225-5252.
Holt, K.E., Y.Y. Teo, et al. 2009. Detecting SNPs and estimating allele frequencies in clonal bacterial populations by sequencing pooled DNA. *Bioinformatics,* 25(16):2074-2075.
Inglesby, T.V., O'Toole, T., Henderson, D.A., Bartlett, J.G., Ascher, M.S., Eitzen, E., et al. 2002. Anthrax as a biological weapon, 2002: Updated recommendations for management. *JAMA,* 287(17):2236-2252.
Jackson, P.J., Walthers, E.A., Kalif, A.S., Richmond, K.L., Adair, D.M., Hill, K.K., et al. 1997. Characterization of the variable-number tandem repeats in vrrA from different *Bacillus anthracis* isolates. *Appl Environ Microbiol,* 63(4):1400-1405.
Jackson, P.J., Hugh-Jones, M.E., et al. 1998. PCR analysis of tissue samples from the 1979 Sverdlovsk anthrax victims: The presence of multiple *Bacillus anthrac*is strains in different victims. *Proc Natl Acad Sci USA,* 95(3):1224-1229.

Jernigan, D.B., Raghunathan, P.L., Bell, B.P., Brechner, R., Bresnitz, E.A., Butler, J.C., et al. 2002. Investigation of bioterrorism-related anthrax, United States, 2001: Epidemiologic findings. *Emerg Infect Dis,* 8(10):1019-1028.

Johnson, N.L., , Kemp, A.W., and S. Kotz. 2005. *Univariate Discrete Distribution,* 3rd ed. New York: Wiley. Chapter 6.

Johnstone, K., Ellar, D.J., and T.C. Appleton. 1980. Location of metal ions in *Bacillus* megaterium spores by high-resolution electron probe X-ray microanalysis. *FEMS Microbiol Lett,* 7:97-101.

Keim, P. 2002a. Forensic Analysis of Putative Anthrax Samples, Batch E0001. February 18, 2002. (B1M3D2).

Keim, P. 2002b. MLVA-15 Molecular Typing (Evidence Number E0001). April 16, 2002. (B1M3D2).

Keim, P. 2009. Northern Arizona University. Presentation to the committee, September 24.

Keim, P., Klevytska, A.M., Price, L.B., Schupp, J.M., Zinser, G., Smith, K.L., et al. 1999. Molecular diversity in *Bacillus anthracis*. *J Appl Microbiol,* 87(2):215-217.

Keim, P., Pearson, T., and R. Okinaka. 2008. Microbial forensics: DNA fingerprinting of *Bacillus anthracis* (anthrax). *Anal Chem,* 80(13):4791-4799.

Keim, P., Price, L.B., Klevytska, A.M., Smith, K.L., Schupp, J.M., Okinaka, R., et al. 2000. Multiple-locus variable-number tandem repeat analysis reveals genetic relationships within *Bacillus anthracis*. *J Bacteriol,* 182(10):2928-2936.

Kendall, C., and T.B. Coplen. 2001. Distribution of oxygen-18 and deuterium in river waters across the United States. *Hydrolog Proc,* 15(7):1363-1393.

Kolsto, A.B., Tourasse, N.J., and O.A. Okstad. 2009. What sets *Bacillus anthracis* apart from other *Bacillus* species? *Annu Rev Microbiol,* 63:451-476.

Kreuzer-Martin, H.W., Chesson, L.A., Lott, M.J., and J.R. Ehleringer. 2005. Stable isotope ratios as a tool in microbial forensics—part 3. Effect of culturing on agar-containing growth media. *J Forensic Sci,* 50(6):1372-1379.

Kreuzer-Martin, H.W., and K.H. Jarman. 2007. Stable isotope ratios and forensic analysis of microorganisms. *Appl Environ Microbiol,* 73(12):3896-3908.

Kreuzer-Martin, H.W., Lott, M.J., Dorigan, J., and J.R. Ehleringer. 2003. Microbe forensics: Oxygen and hydrogen stable isotope ratios in *Bacillus subtilis* cells and spores. *Proc Natl Acad Sci USA,* 100(3):815-819.

Kuhlman, M.R. 2001a. Preliminary SPOT Report on Particle Size Analyses. Battelle Memorial Institute. October 18. (B2M13D2).

Kuhlman, M.R. 2001b. Preliminary SPOT Report on Sample Analyses. Battelle Memorial Institute. October 22. (B2M13D5).

Kuhlman, M.R. 2001c. SPOT Report on Analyses of Silicon and Silica in Powder Samples: SEM/EDS Analysis. Battelle Memorial Institute. November 26. (B2M13D8).

Kunst, F., Ogasawara, N., Moszer, I., Albertini, A.M., Alloni, G., Azevedo, V., et al. 1997. The complete genome sequence of the Gram-positive bacterium *Bacillus subtilis*. *Nature,* 390(6657):249-256.

Leffel, E.K., and L.M. Pitt. 2006. Anthrax. In J.R. Swearengen (ed.), *Biodefense: Research Methodology and Animal Models* (pp. 77-93). Boca Raton, FL: Taylor & Francis.

Levin, I., and B. Kromer. 2004. The tropospheric $^{14}CO_2$ level in mid-latitudes of the Northern Hemisphere (1959-2003). *Radiocarbon,* 46(3):1261-1272.

Liddington, R.C. 2002. Anthrax: A molecular full nelson. *Nature,* 415(6870):373-374.

Malakoff, D. 2002. Bioterrorism. Student charged with possessing anthrax. *Science,* 297(5582):751-752.

Malorny, B., Cook, N., D'Agostino, M., DeMedici, D., Croci., L., et al. **2004. Multicenter valida**tion of PCR-based method for detection of *Salmonella* in chicken and pig samples. *J AOAC Int,* 87:861-866.

Mann, S., Sparks, N.H., Scott, G.H., and E.W. de Vrind-de Jong. 1988. Oxidation of manganese and formation of Mn(3)O(4) (Hausmannite) by spore coats of a marine *Bacillus sp*. *Appl Environ Microbiol,* 54(8):2140-2143.

Martin, D. 2010. Dugway Proving Ground. Presentation to the committee, April 23.
Meselson, M., Guillemin, J., et al. 1994. The Sverdlovsk anthrax outbreak of 1979. *Science*, 266(5188):1202-1208.
Michael, J. 2009. Sandia National Laboratory. Presentation to the committee, September 25.
Michel, J.F., Cami, B., and P. Schaeffer. 1968. (Selection of *Bacillus subtilis* mutants blocked at the beginning of sporulation. II. Selection by adaptation to a new carbon source and by aging of the sporulated cultures). *Ann Inst Pasteur (Paris)*, 114(1):21-27.
Mock, M., and A. Fouet. 2001. Anthrax. *Annu Rev Microbiol*, 55:647-671.
Moir, A. 2006. How do spores germinate? *J Appl Microbiol*, 101(3):526-530.
NRC (National Research Council). 2009a. *Strengthening Forensic Science in the United States: A Path Forward*. Washington, DC: The National Academies Press.
NRC 2009b. *Responsible Research with Biological Select Agents and Toxins*. Washington, DC: The National Academies Press.
New York Times. 2002. A Nation Challenged: Anthrax; Senate Offices Refumigated a Second Time. (January 1).
Nicholson, W.L. and P. Setlow 1990. Sporulation, germination, and outgrowth. In *Molecular Biological Methods for* Bacillus (pp. 391-450). C.R. Harwood and S.M. Cutting (eds.) Chichester: Wiley.
Nicholson, W.L., Munakata, N., Horneck, G., Melosh, H.J., and P. Setlow. 2000. Resistance of *Bacillus* endospores to extreme terrestrial and extraterrestrial environments. *Microbiol Mol Biol Rev*, 64(3):548-572.
Pacific Northwest National Laboratory. (Various Dates). Analyses for Detection of Residual Agar. (B1M11).
Pesenti P. "A U.S. research strategy for microbial forensics: from genesis to implementation". In: Microbial Forensics, 2nd edition, eds, Budowle B., Schutzer S.E., Breeze R., Keim P.S., Morse S.A.. Academic Press, Amsterdam, 2010, pp. 605-617.
Piggee, C. 2008. Tracing Killer Spores. *Anal Chem*, September 18. Online News http://pubs3.acs.org/journals/ancham/news/2008/09/18/cp_anthrax.html.
Pomerantsev, A.P., Staritsin, N.A., Mockov Yu, V., and L.I. Marinin. 1997. Expression of cereolysine AB genes in *Bacillus anthracis* vaccine strain ensures protection against experimental hemolytic anthrax infection. *Vaccine*, 15(17-18):1846-1850.
Price, L.B., Hugh-Jones, M., Jackson, P.J., and P. Keim. 1999. Genetic diversity in the protective antigen gene of *Bacillus anthracis*. *J Bacteriol*, 181(8):2358-2362.
Okinaka, R.T., Henrie, M., Hill, et al. 2008. Single nucleotide polymorphism typing of *Bacillus anthracis* from Sverdlovsk tissue. *Emerg Infect Dis*, 14(4):653-656.
Ravel, J. 2009. The Genomics behind the Amerithrax Investigation. Presentation before the Biodefense Meeting of the American Society for Microbiology. February 25, 2009.
Ravel, J., Jiang, L., Stanley, S.T., Wilson, M.R., Decker, R.S., Read, T.D., et al. 2009. The complete genome sequence of *Bacillus anthracis* Ames "Ancestor." *J Bacteriol*, 191(1):445-446.
Ravel, J. 2010. Communication to the committee, May 13.
Read, T.D., Peterson, S.N., Tourasse, N., Baillie, L.W., Paulsen, I.T., Nelson, K.E., et al. 2003. The genome sequence of *Bacillus anthracis* Ames and comparison to closely related bacteria. *Nature*, 423(6935):81-86.
Read, T.D., Salzberg, S.L., Pop, M., Shumway, M., Umayam, L., Jiang, L., et al. 2002. Comparative genome sequencing for discovery of novel polymorphisms in *Bacillus anthracis*. *Science*, 296(5575):2028-2033.
Sanderson, W., Stoddard, R., Echt, A., McCleery, R.E., Picitelli, C.A., Kim, D., et al. 2001. Evaluation of *Bacillus anthracis* contamination inside the Brentwood Post Office, Washington, D.C. Report to U.S. Postal Service. Cincinnati, OH: National Institute for Occupational Safety and Health.

Sanderson, W., Hein, M., Taylor, L., Curwin, B., Kinnes, G., Hales, T., et al. 2002. Second evaluation of *Bacillus anthracis* contamination inside the Brentwood Mail Processing and Distribution Center, District of Columbia. Report to the U.S. Postal Service. Cincinnati, OH: National Institute for Occupational Safety and Health.

Sarmiento, G. 2007. Former AMI Building Declared Free of Anthrax Contamination. *Palm Beach Post*. February 8.

Sand, L., et al. 2009. *Modern Federal Jury Instructions*. Albany, NY: Matthew Bender.

Sastalla, I., Rosovitz, M.J., and Leppla, S.H. 2010. Accidental Selection and Intentional Restoration of Sporulation-Deficient *Bacillus anthracis* Mutants. *Appl Environ Microbiol,* 76(18):6318-6231.

Schutzer, S. 2009. University of Medicine and Dentistry of New Jersey. Presentation to the committee, September 24.

Setlow, P. 2003. Spore germination. *Curr Opin Microbiol,* 6(6):550-556.

Setlow, P. 2006. Spores of *Bacillus subtilis*: Their resistance to and killing by radiation, heat and chemicals. *J Appl Microbiol,* 101(3):514-525.

Smith, J. 2009. BIOFOR Consulting. Presentation to the committee, July 31.

Snedecor, G.W., and Cochran, W.G. 1989. *Statistical methods* (8th ed.). Ames: Iowa State University Press.

Stewart, M., Somlyo, A.P., Somlyo, A.V., Shuman, H., Lindsay, J.A., and W.G. Murrell. 1980. Distribution of calcium and other elements in cryosectioned *Bacillus cereus* T spores, determined by high-resolution scanning electron probe X-ray microanalysis. *J Bacteriol,* 143(1):481-491.

Stewart, M., Somlyo, A.P., Somlyo, A.V., Shuman, H., Lindsay, J.A., and W.G. Murrell. 1981. Scanning electron probe X-ray microanalysis of elemental distributions in freeze-dried cryosections of *Bacillus coagulans* spores. *J Bacteriol,* 147(2):670-674.

Swartz, M.N. 2001. Recognition and management of anthrax: An update. *N Engl J Med,* 345:1621.

Tamir, H., and C. Gilvarg. 1966. Density gradient centrifugation for the separation of sporulating forms of bacteria. *J Biol Chem,* 241(5):1085-1090.

Temple-Raston, D. 2008, September 17. Mueller: FBI Needs More Powers to Combat Threats. National Public Radio.

Teshale, E.H., Painter, J., et al. 2002. Environmental sampling for spores of *Bacillus anthracis. Emerg Infect Dis,* 8(10):1083-1087.

Traeger MS et al. 2002. First case of bioterrorism-related inhalational anthrax in the United States, Palm Beach County, Florida, 2001. *Emerg Infect Dis,* 8:1029-1034.

USAMRIID (United States Army Medical Research Institute for Infectious Diseases). 1997. Reference Material Receipt Record, October 22. Retrieved September 2, 2010, from http://www.anthraxinvestigation.com/usamriid-rmrr-bldg-1412-p-1-of-1.jpg.

USAMRIID. 2001. EM Daschle "Si" Report. October 25. (B1M2D4).

USAMRIID. 2004. "USAMRIID Highlights." April 27. Retrieved August 4, 2010, from http://www.usamriid.army.mil/highlightspage.htm.

USDOJ (United States Department of Justice). 2010. *Amerithrax Investigative Summary*. February 19, 2010, updated October 15, 2010. Available at: http://www.justice.gov/amerithrax/docs/amx-investigative-summary.pdf.

USDOJ. 2000. *Crime scene investigation: A guide for law enforcement. Research Report*. Office of Justice Programs. National Institute of Justice.

University of Maryland, Battelle Memorial Institute, and Edgewood Chemical Biological Center. (Various dates). Agar and Heme Analysis. (B1M10).

Van Ert, M.N., Easterday, W.R., Huynh, L.Y., Okinaka, R.T., Hugh-Jones, M E., Ravel, J., et al. 2007a. Global genetic population structure of *Bacillus anthracis*. *PLoS One,* 2(5):e461.

Van Ert, M.N., Easterday, W.R., Simonson, T.S., U'Ren, J.M., Pearson, T., Kenefic, L.J., et al. 2007b. Strain-specific single-nucleotide polymorphism assays for the *Bacillus anthracis* Ames strain. *J Clin Microbiol,* 45(1):47-53.

Velicer, G.J., Kroos, L., and R.E. Lenski. 1998. Loss of social behaviors by *Myxococcus xanthus* during evolution in an unstructured habitat. *Proc Natl Acad Sci USA,* 95(21):12376-12380.

Wahl, K.L., Colburn, H.A., Wunschel, D.S., Petersen, C.E., Jarman, K.H., and N.B. Valentine. 2010. Residual agar determination in bacterial spores by electrospray ionization mass spectrometry. *Anal Chem,* 82(4):1200-1206.

Washington Post. 2008. "Hatfill Timeline," August 9, p. 11.

Weber, P. 2009. Lawrence Livermore National Laboratory. Presentation to the committee, September 25.

Whiteaker, J.R., Fenselau, C.C., Fetterolf, D., Steele, D., and D. Wilson. 2004. Quantitative determination of heme for forensic characterization of *Bacillus* spores using matrix-assisted laser desorption/ionization time-of-flight mass spectrometry. *Anal Chem,* 76(10):2836-2841.

Wilkening, D.A. 2006. Sverdlovsk revisited: Modeling human inhalation anthrax. *Proc Natl Acad Sci USA,* 103(20):7589-7594.

Wilkening, D.A. 2008. Modeling the incubation period of inhalational anthrax. *Med Decis Making,* 28(4):593-605.

Wilson, D. 2009. Federal Bureau of Investigation. Presentation to the committee, December 10.

Worsham, P. 2009. United States Army Medical Research Institute for Infectious Diseases. Presentation to the committee, September 24.

Worsham, P.L., and M.R. Sowers. 1999. Isolation of an asporogenic (*spo0A*) protective antigen-producing strain of *Bacillus anthracis. Can J Microbiol,* 45(1):1-8.

Wunschel, D.S., Colburn, H.A., Fox, A., Fox, K.F., Harley, W.M., Wahl, J.H., et al. 2008. Detection of agar, by analysis of sugar markers, associated with *Bacillus anthracis* spores, after culture. *J Microbiol Meth,* 74(2-3):57-63.

Yamada, S., Ohashi, E., Agata, N., and K. Venkateswaran. 1999. Cloning and nucleotide sequence analysis of gyrB of *Bacillus cereus, B. thuringiensis, B. mycoides,* and *B. anthracis* and their application to the detection of *B. cereus* in rice. *Appl Environ Microbiol,* 65:1483-1490.

Index of Documents Provided by the Federal Bureau of Investigation[1]

BATCH 1

Module 1: Technical Review Panel Meetings

Technical Review Panels

1. November 5, 2001 WFO Anthrax Expert Panel Review (pp. 2-27)
2. December 7, 2001 External Technical Review Panel (pp. 28-39)
3. December 12, 2001 External Review of Analytical Plan (pp. 40-46)
4. June 11-12, 2002 Progress to Date Review (pp. 47-65)
5. August 8, 2005 Chemistry Review Panel (pp. 66-195)

Module 2: USAMRIID

Initial examination of the letter spore preparations for physical characteristics (microscopy and electron microscopy) and spore viability studies.

1. 18 Oct 2001 SPS02.57 (Daschle) CFU Report (p. 3)
2. 21 Oct 2001 EM Report of Dasch1e Letter (pp. 4-19)

---

[1] The Committee on the Review of the Scientific Approaches Used During the FBI's Investigation of the 2001 *Bacillus anthracis* Mailings initially received FBI materials related to the science of the Anthrax Mailings Investigation in two batches. A third batch of documents was provided later. In addition to the documents in the three principal batches of materials, the FBI also provided several supplemental documents in response to committee requests. Unless otherwise noted, this index reflects the order in which the materials were received from the FBI, the manner in which the documents were organized by the FBI, and the language used by the FBI when describing the documents. Page numbers in parentheses after document titles are the page numbers assigned to the document by the FBI or, in cases where the FBI did not assign page numbers, the number of pages in the respective document.

3. 24 Oct 2001 SPS02.88.1 (*NY Post*) CFU Report (pp. 20-21)
4. 25 Oct 2001 EM Daschle "Si" Report (pp. 22-25)
5. 25 Oct 2001 Simons Letter re *NY Post* (pp. 26-28)
6. 28 Oct 2001 EM/CFU Report of *NY Post* (p. 29)
7. 27 Nov 2001 Report on Isolates from Daschle and *NY Post* Letters (p. 30)

Special Pathogens Sample Test Laboratory Analytical Test Reports: Results of Analysis of Letter Material

8. SPS.02.88 *NY Post* Powder 10/22/2001 (pp. 32-33)
9. SPS.02.44 Brokaw Envelope 11/4/2001 (pp. 34-35)
10. SPS.02.57 Daschle Letters and Powders 11/9/2001 (pp. 36-45)
11. SPS.02.266 Leahy Powder (pp. 46-53)

Microbiological examinations and identification of phenotypic variants (Morphotypes) which appeared different than the predominant "Ames" phenotype

12. Report #1 Analysis of Evidentiary Material (pp. 55-68)
13. Report #2 Analysis of Repository Samples (pp. 69-79)
14. Report #3 Analysis of Environmental Samples - AMI (pp. 80-90)
15. Report #4 Examination of Repository Spore Preparations: Screening for the Hemolytic *B. subtilis* Contaminant (pp. 91-95)
16. Report #5 Analysis of Repository Samples (pp. 96-99)

Isolation of Morphological Variants from FBI Repository Samples FBIR 049 004 (Leahy) PowerPoint Photos

17. Isolation of Morphological Variants from FBI Repository Samples FBIR 049 004 (Leahy) PowerPoint Photos (pp. 101-109)

Module 3: Ames Strain Identification

Reports

1. Forensic Analysis of Putative Anthrax Samples, Batch E0001 v02.01.02 (MLVA-8) (pp. 3-10)
2. MLVA-15 Molecular Typing (Batch E0001) 4/16/02 (pp. 11-14)
3. Laboratory Reports NAU-0001 (May 30, 2002) to NAU-0038 (May 5, 2008) (pp. 15-167)

Guidelines and Standard Operating Procedures

> 4. Guidelines and Standard Operating Procedures for Forensic Analysis V09.23.03 (pp. 168-204)
> 5. Guidelines and Standard Operating Procedures for Forensic Analysis: Real-Time PCR species Specific, Canonical and Strain Specific SNP Genotyping of *Bacillus anthracis* and *Francisella tularensis* V11.03.04 (pp. 206-231)
> 6. Real-Time PCR species Specific, Canonical and Strain Specific SNP Genotyping of *Bacillus anthracis* V04.11.07 (pp. 232-272)
> 7. Quality Assurance Standards for Forensic DNA Analysis (pp. 273-280)

Literature

> 8. "Multiple-Locus Variable-Number Tandem Repeat Analysis Reveals Genetic Relationships within *Bacillus anthracis*" (Keim et al., 2000, 182:2928-2936) (pp. 282-291)

Progress Reports

> 9. NAU DNA-Based Strain Typing of Anthrax Samples (pp. 293-649)[2]

Module 4: Analysis for Evidence of Genetic Engineering

Reports

> 1. 19 October 2001 LANL Receipt letter (pp. 3-4)
> 2. Laboratory Reports (pp. 5-55)

Presentation

> 3. Analysis of the Amerithrax *B. Anthracis* Ames Isolates for Evidence of Genetic Engineering (pp. 57-79)

Module 5: Genomic Sequencing

Publications

> 1. "Global Genetic Population Structure of *Bacillus anthracis*," M. Van Ert et al., (2007) PLoS ONE 2(5); e461. doil-. 1371/jpornal.pone.0000461 (pp. 2-11)

---

[2] The FBI did not assign a name to this document. The descriptor was assigned by the committee.

2. "The Complete Genome Sequence of *Bacillus anthracis* Ames 'Ancestor,'" J. Ravel et al, *J. Bacteriology*, Jan 2009, Vol 191, No. 1, p. 445-446 (pp. 12-13)

Reports

3. Genomic Analysis of *Bacillus anthracis* Isolates Relevant to the Amerithrax Investigation, June 1, 2004 (Morph A, B, C, D, Wild type) (pp. 17-75)
4. Genomic Analysis of *Bacillus anthracis* Isolates Relevant to the Amerithrax Investigation, June 4, 2005 (Morph E (Opaque, *Post/Leahy B.s.*) (pp. 76-159)
5. Multiple Locus PCR-based Assay for the Direct Comparison of unknown *B. subtilis* isolates to *B. Subtilis, New York Post*, May 15, 2006 (pp. 160-220)

TIGR Progress Reports

6. TIGR Progress Reports (pp. 221-458)

Module 6: Sandia National Laboratory (SNL)

Reports

1. Sandia National Laboratory Final Report (pp. 3-42)

Historical Literature

2. "Distribution of Calcium and Other Elements in Cryosectioned *Bacillus cereus T* Spores, determined by High-Resolution Scanning Electron Probe X-Ray Microanalysis," M. Stewart, A.P. Somlyo, A.V. Somlyo, H. Shuman, J.A. Lindsay, W.G. Murrell, *J .Bacteriology,* July 1980, Vol 143, No. 1, pp. 481-491 (pp. 44-54)
3. "Scanning Electron Probe X-Ray Microanalysis of Elemental Distribution in Freeze-Dried Cryosections of *Bacillus coagulans* Spores." M. Stewart, A.P. Somlyo, A.V. Somlyo, H. Shuman, J.A. Lindsay, W.G. Murrell, *J . Bacteriology*, Aug. 1981, Vol 147, No. 2, pp. 670-674 (pp. 55-59)
4. "Automated Analysis of SEM X-Ray Spectral Images: A Powerful New Microanalysis Tool," P.G. Kotula, M.R. Kennan, J.R. Michael, *Microsc. Microanal.,* 9, 1-17, 2003 (pp. 60-76)
5. Silicon Summary Bibliography of Select Publications (pp. 77-81)

INDEX OF DOCUMENTS PROVIDED BY THE FBI

PowerPoint Files of Individual Sample Results

6. PowerPoint Files of Individual Sample Results (pp. 82-332)

Module 7: Chemistry Unit, FBI Laboratory

FBI Laboratory Reports

1. 020322006 April 15, 2002 — Q13 SPS02.88 (Post) (pp. 3-4)
2. 020605001 June 18, 2002 — Q12 White Powder (Leahy) (pp. 5-6)
3. 020110004 August 26, 2002 — Q12, Bc 14579, Bs/Ba+/ (pp. 7-10)
4. 020605001 October 16, 2002 — Q13 SPS02.88 Elemental (pp. 11-12)
5. 040624018 June 28, 2004 — Leahy Powder Elemental (pp. 13-14)
6. 030519001 October 15, 2003 — Culture Media Elemental (pp. 15-22)
7. 031008001 October 30, 2003 — Burans 12 Elemental (pp. 23-25)
8. 031124029 December 11, 2003 — 10 DPG growths Elemental (pp. 26-28)
9. 030819015 June 29, 2004 — Commercial Media (pp. 29-34)
10. 050321004 April 18, 2005 — Media Salts and Spores (pp. 35-37)
11. 050118006 April 18, 2005 — 19 DPG Stubs SEM (pp. 38-54)
12. 050408005 May 18, 2005 — 2 DPG Stubs SEM (pp. 55-67)
13. 050408005 July 6, 2005 — 2 DPG Stubs SEM (pp. 68-91)

Elemental Analysis Summary

14. 1 Elemental Analysis Summary Table (pp. 93-94)
15. Comparison of results from 2 different instruments (by 2 different FBI examiners, 2 years apart in 2 different Labs (HQ VS Quantico)

Envelopes and Particle Transport

16. FBI Laboratory Electron Microscopy of Envelope surfaces and particle transport thru envelopes (pp. 96-111)

Module 8: Carbon-14 (^{14}C) Dating

Center for Accelerator Mass Spectrometry (CAMS), Lawrence Livermore National Laboratory

1. CAMS background information (pp. 3-14)
2. Proposal (pp. 15-16)
3. Quantitating Radiocarbon Concentrations in Isolated Samples of Biological Origin: Standard Operating Procedures for FBI Measurements of Natural 14C (pp. 17-23)
4. LLNL Report 10/14/02 (pp. 24-29)
5. LLNL Report 1/16/04 Addendum (pp. 30-33)
6. LLNL Data Table 3/30/04 DPG Samples (pp. 34-39)
7. Scientific References (pp. 40-50)

National Ocean Sciences Accelerator Mass Spectrometry Facility (NOSAMS), Woods Hole Oceanographic Institute

8. NOSAMS background information (pp. 52-55)
9. General Statement of C-14 Procedures (pp. 56-57)
10. Amerithrax Sample Handling Procedures (pp. 58-72)
11. Final Report (pp. 73-83)
12. Scientific References (pp. 84-112)

Module 9: Stable Isotopes

Stable Isotope Reports

1. Isotopic Characterization of Water Samples 3/14/03 (pp. 3-10)
2. Isotopic Characterization of Anthrax Samples 7/30/03 (pp. 11-19)
3. Isotopic Characterization of Water Samples 10/15/03 (pp. 20-25)
4. Isotopic Characterization of Spore Samples 1/5/04 (pp. 26-31)
5. Stable Isotope Characterization of Anthrax Sample SPS 02.266 2/22/04 (pp. 32-44)
6. Isotopic Characterization of Microbial Spore Samples 5/20/04 (pp. 45-53)
7. A Report on the Stable Isotope Ratios of Envelope Samples 9/17/04 (pp. 54-58)
8. A Report on the Stable Isotope Ratios of Treated and Untreated Envelopes 11/7/04 (pp. 59-64)
9. A Report on the Stable Isotope Ratios of Treated and Untreated Envelopes 12/22/04 (pp. 65-77)
10. Isotopic Characterization of RMR 1029 5/6/05 (pp. 78-80)
11. A Tabulation of Stable Isotopes of Tap Water Samples Analyzed for the FBI 3/23/05 (pp. 81-247)

Proposals and SOWS

12. Proposals and SOWS (pp. 248-269)

Kreuzer-Martin Publications

13. Kreuzer Martin Publications (pp. 271-310)

Stable Isotope Ratios and the Forensic Analysis of Microorganisms PowerPoint

14. Stable Isotope Ratios and the Forensic Analysis of Microorganisms PowerPoint (pp. 312-342)

Module 10: Agar and Heme Analysis

Agar Assay University of Maryland

1. Unsolicited Proposal SOW (pp. 3-8)
2. Progress Reports (pp. 9-55)
3. Summary of Chromatograms (pp. 56-72)
4. Final Agar Report (pp. 73-94)
5. Reductive Hydrolysis Procedure Steps (pp. 95-96)
6. Reductive Hydrolysis Literature References (pp. 97-118)
7. ASMS Abstract Poster (pp. 119-120)

Agar Assay Validation - BMI

8. Task 4 Spot Report 27 Aug 2002 (pp. 122-126)
9. Task 4 Interim Report 17 Sept 2002 (pp. 127-136)

Heme Assay (UMD)

10. SOW (pp. 138-149)
11. Progress Reports (pp. 150-182)
12. Analysis of Heme by MALDI Procedure (pp. 183-190)
13. Filter Sterilization Validation (pp. 191-192)
14. Heme Final Report (pp. 193-216)
15. Heme A.C. publication (pp. 217-222)

Heme Assay Validation – ECBC

16. Interim Report (ECBC) (pp. 224-238)
17. Analytical Test Report (ECBC) (10 Aug 2005) (pp. 239-246)

Agar Overview

18. Agar Overview (pp. 248-251)

Module 11: Pacific Northwest National Laboratory

Reports

1. Residual Agar Determination in Bacterial Cultures by ESI/MS, DHS Final Report (pp. 3-99)
2. Detection of Agar, by Analysis of Sugar Markers, Wunschel et al., *J. Microbiol Methods,* 74, (2008), 57-63 (pp. 101-107)
3. Residual Agar Determination in Bacterial Cultures by ESI/MS, K.L. Wahl et al. submitted for publication (pp. 108-141)

Presentation

4. FBI Samples Data Summary PowerPoint (pp. 143-197)

Module 12: FBI Laboratory Renocal Assay

Reports

1. FBI Lab Report 070829018 (pp. 2-3)

Standard Operating Procedures

2. Meglumine Diatrizoate Analysis by LC/MS/ESI (pp. 5-26)
3. Performance Monitoring Protocol (QA/QC) for the Finnigan LTQ LC/MS (ESI) Instrument (pp. 27-45)

Reference

4. Detection of Trace Amounts of Meglumine and Diatizoate from *Bacillus* Spore Samples Using Liquid Chromatography/Mass Spectrometry; submitted to *Anal. Chem.* (pp. 47-74)

Module 13: DPG Production Methods

Test Plan

1. Test Plan (pp. 3-24)

Project Overview PowerPoint

2. Project Overview PowerPoint (pp. 26-60)

Final Report Dated 2/1/06

3. Final Report Dated 2/1/06 (pp. 67-118)

SEM Photos of Production

4. SEM Photos of Production (pp. 120-255)

Module 14: Preparation of *B. anthracis* Ames on Commercial and Prepared Media

Report

1. Preparation of *B. anthracis* Ames Spores on Commercially Prepared Media and Media Prepared at USAMRIID (pp. 3-6)

BATCH 2

Module 1: *B. Subtilis* Contaminant

Battelle

1. Report of isolation of contaminant from Brokaw letter in "Summary of Microbiological Analysis 19 Oct 2001" (pp. 3-5)

Centers for Disease Control

2. Report of identification of contaminant submitted by BMI as *B. subtilis* including morphology, hemolysis, gram staining, antibiotic resistance and 16s rRNA sequence match (pp. 7-19)

Novazymes

3. Identification of the *NY Post* contaminant as *B. licheniformis* (p. 21)

Applied Biosystems

4. Identification of the *B. subtilis* isolated from the Brokaw and *NY Post* letters by 16s ribosomal RNA gene sequence analysis (pp. 23-32)

TIGR

> Genomic Analysis of *Bacillus anthracis* Isolates Relevant to the Amerithrax Investigation, June 4, 2005 (Morph E (Opaque), *Post/Leahy B.s.*) (see Batch 1, Module 5, Document 3)
>
> Multiple Locus PCR-based Assay for the Direct Comparison of Unknown *B. subtilis* Isolates to *B. subtilis, New York Post,* May 15, 2006 (see Batch 1, Module 5, Document 4)

Module 2: Whole Genome Assembly of *B. subtilis* Isolate

Technical Proposals and SOW

1. Microbial Genetic Services in Support of the Amerithrax Investigation (FBI SOW) (pp. 3-42)
2. Whole Genome Assembly, Closure Annotation of *Bacillus subtilis* GB22 (TO1) (pp. 43-57)
3. Multiple Locus PCR-Based Assay of the Direct Comparison of One Unknown *B. subtilis* Isolate to *B. subtilis New York Post* (TO3) (pp. 58-74)
4. Training for Use of Affymetrix Comparative Genomic Hybridization Arrays (TO4) (pp. 75-91)
5. Develop an Annotation File Specific to the FBI *B. Subtilis* Comparative Genomic Hybridization Arrays (New TO5) (pp. 92-101)

Progress Reports

6. Whole Genome Assembly, Closure and Annotation of *Bacillus subtilis* GB2212/07-4/08 (TO1) (pp. 103-106)
7. Multiple Locus PCR-based Assay of the Direct Comparison of One Unknown *B. subtilis* Isolate to *B. subtilis New York Post* 1/08-4/08 (TO3) (pp. 107-114)
8. Multiple locus PCR-Based Assay of the Direct Comparison of One Unknown *B. subtilis* Isolate to *B. subtilis New York Post* 5/08 (TO3) (pp. 115-122)

Standard Operating Procedures

9. Affymetrix GB22 Tiling GeneChip SOP (pp. 124-136)

Final Reports

> 10. Whole Genome Assembly, Closure and Annotation of *Bacillus subtilis* GB22 (TO1) (pp. 138-142)
> 11. Multiple Locus PCR-Based Assay of the Direct Comparison of One Unknown *B. subtilis* Isolate to *B. subtilis New York Post* (TO3) (pp. 143-166)
> 12. Genome MTV Manual (TO4) (pp. 167-185)

Module 3: Genetic Diversity and Phylogenetic Characterization of *B. subtilis*

Statement of Work (SOW)

> 1. Genetic Diversity and Phylogenetic Characterization of *B. subtilis* (pp. 3-14)

Progress Reports

> 2. 8 Monthly and 1 Quarterly Report 12/06-11/07 (pp. 16-79)

Standard Operating Procedures

> 3. Allele-Specific Oligonucleotide (ASO) Typing Assay for *Bacillus subtilis* (pp. 81-82)

Final Reports

> 4. Genetic Diversity and Phylogenetic Characterization of *B. subtilis* 11/30/07 (pp. 84-124)

Module 4: *B. subtilis* Screening

Standard Operating Procedures

> 1. *Bacillus subtilis* Analysis by Singleplex Real-Time PCR Standard Operating Procedure (pp. 3-10)

Synopsis of Assay Development

> 2. Synopsis of Assay Development (pp. 11-55)

Standard Operating Procedures

    3. Analytical Plan, Acceptance of Work (AOW) 40 (pp. 57-104)

Analytical Plans

    4. Analytical Plan, Acceptance of Work (AOW) 46 (pp. 106-115)
    5. Analytical Plan, Acceptance of Work (AOW) 47 (pp. 116-183)
    6. Analytical Plan, Acceptance of Work (AOW) 57 (pp.184-187)
    7. Analytical Plan, Acceptance of Work (AOW) 58 (pp. 188-193)
    8. Analytical Plan, Acceptance of Work (AOW) 62 (pp. 194-239)
    9. Analytical Plan, Acceptance of Work (AOW) 68 (pp. 240-244)

Assay Validation

    10. Serial Dilution/LOD   NBFAC.061117.001 (pp. 246-253)
    11. Assay Validation   NBFAC.061019.002 (pp. 254-259)

Laboratory Reports

    12. Location Searches   NBFAC. 061113.001 (pp. 261-279)
    13. Repository Screening   NBFAC.070215.0001,
                                    NBFAC.070314.0001
                                    to NBFAC0.70314.003 (pp. 280-359)
    14. Environmental Samples   NBFAC.070723.0001 (pp. 360-380)
    15. Location Searches   NBFAC.070727.0001,
                                    NBFAC.071102.0001,
                                    NBFAC.080828.0001 (pp. 381-440)

Module 5: Molecular Analysis of Pathogen Strains and Isolates and Genetic Mutations A1 and A3

Technical Proposal and FBI SOWs: Genetic Discrimination of *Bacillus anthracis* Isolates Using Molecular Biological Techniques

    1. Technical Proposal and FBI SOWs: Genetic Discrimination of *Bacillus anthracis* Isolates Using Molecular Biological Techniques (pp. 3-65)

Assay Development Morph A1-A3

    2. Progress Reports: May 2002-Jan 2004 (pp. 67-499)

Assay Development Morph A1-A3

    3. Final Report (pp. 501-603)
    4. Validation Study Morph A-1 Protocol (pp. 604-712)
    5. Validation Study Morph A-2 Protocol (pp. 713-820)
    6. Validation Study Morph A-3 Protocol (pp. 821-932)

Assay Development Morph A1-A3

    7. Morph A Overview PowerPoint (pp. 934-957)

Repository Screening Morph A1 and A3

    8. Progress and Final Reports, March 2004-Oct 2007 (pp. 959-1295)

Module 6: Genetic Mutations B and D (CBI)

Assay Development: Technical Proposals

1. DNA Assays for Minor Genetic Variants of *Bacillus anthracis* (pp. 3-76)
2. Task 1 Technical Proposal Morph B SNP (pp. 77-94)
3. Task 2 Technical Proposal Morph D Deletion (pp. 95-114)

Assay Development: Progress Reports

4. Morph B (pp. 116-160)
5. Morph D (pp. 161-222)

Assay Development: Standard Operating Procedures

6. Morph B Protocol (pp. 224-268)
7. Morph D Protocol (pp. 269-310)
8. Appendix 1-10 (pp. 311-867)

Module 7: Genetic Mutations B and D (IITRI)

Assay Development: Technical Proposals

1. Technical Proposal (pp. 3-51)
2. Task 1 DNA Assay Development – Morph B SNP (pp. 52-97)
3. Task 2 DNA Assay Development – Morph D Deletion (pp. 98-137)
4. Capabilities Brief (pp. 138-176)

174   SCIENTIFIC APPROACHES USED TO INVESTIGATE THE ANTHRAX LETTERS

Assay Development: Progress Reports

    5. DNA Assays for Minor Genetic Variants Meeting Minutes 2/23/05 (pp. 178-179)
    6. Task 1 Morph B SNP Progress Reports (pp. 180-215)
    7. Task 2 Morph D Progress Reports (pp. 216-251)

Assay Development: Validation

    8. Validity Test Reports B SNP (pp. 253-408)
    9. Validity Test Report D Deletion (pp. 409-661)

Assay Development: Standard Operating Procedures

    10. Standard Operating Procedure Real-Time PCR Assay for the Detection of the Morph B SNP (pp. 662-686)
    11. Standard Operating Procedure Real-Time PCR Assay for the Detection of the Morph D Deletion (pp. 687-710)

Repository Screening Morph D: Progress Reports

    12. Monthly Status Reports (pp. 711-800 and 803-955)
    13. FBI Kick-Off Meeting (pp. 956-996)
    14. *B. anthracis* Morph D DNA Screening Progress Reviews (pp. 997-1016)

Repository Screening Morph D: Final Reports

    15. Technical Repository Screening Final Reports (pp. 1018-1134)
    16. Option Period I, II, III Final Reports (pp. 1135-1212)

Module 8: Genetic Mutations B and D (MRI)

Assay Development: Technical Proposals

    1. Volume 1 Technical Proposal (pp. 3-57)
    2. Task 1 DNA Assay Development – Morph B SNP (pp. 58-80)
    3. Task 2 DNA Assay Development – Morph D Deletion (pp. 81-102)

Assay Development: Progress Reports

    4. Task 1 Morph B SNP Progress Reports (pp. 104-170)
    5. Task 2 Morph D Deletion Progress Reports (pp. 171-242)

Assay Development: Validation

   6. Morph B SNP Assay Validity Test Report (pp. 244-260)
   7. Morph D Deletion Assay Validity Test Report (pp. 261-272)

Assay Development: Standard Operating Procedures

   8. Extraction of DNA from *Bacillus anthracis* Cultures (pp. 274-281)
   9. PCR-Based Assay for the Detection of Morph B SNP (pp. 282-315)
   10. PCR-Based Assay for the Detection of Morph D Deletion (pp. 316-350)

Assay Development: Final Reports

   11. Task 1 Morph B Final Administrative Report (pp. 352-358)
   12. Task 2 Morph D Final Administrative Report (pp. 359-366)
   13. DNA Assay Development Morph B SNP (pp. 367-380)
   14. DNA Assay Development Morph D Deletion (pp. 381-396)

Repository Screening Morph D: Technical Proposals

   15. Volume 1 Technical Proposal (pp. 398-435)

Repository Screening Morph D: Progress Reports

   16. DNA Screening of Ames Strain Anthrax Samples for Morph D (pp. 437-628)

Repository Screening Morph D: Final Reports

   17. Technical Repository Screening Report (pp. 630-670)
   18. Technical Repository Screening Report Addendum 1 (pp. 671-675)
   19. Technical Repository Screening Report Addendum 2 (pp. 676-680)

Module 9: Genetic Mutation E (TIGR)

Assay Development: Technical Proposals

   1. LL19 Detection Assay: Validation Study Analysis Plan (pp. 3-7)
   2. LL19 Detection Assay: Analysis Plan for 1060 Blind Samples (pp. 8-13)

Assay Development: Progress Reports

    3. June 2004 thru Sept/Dec 2005 (see Batch 1, Module 5)

Assay Development: Standard Operating Procedure

    4. Opaque Assay Development SOP (pp. 16-58)

Assay Development: Validation

    5. *B. anthracis* LL19 Detection Assay: Validation Study (pp. 60-123)

Assay Development: Final Report

    6. Genomic Analysis of *Bacillus anthracis* Isolates Relevant to the Amerithrax Investigation, June 4, 2005 (Morph E (Opaque, *Post/Leahy B.s.*) (see Batch 1, Module 5)

Repository Screening Morph E: Final Report

    7. Analysis of a Repository of *Bacillus anthracis* for the Presence of the LL19 Opaque Deletion Genotype (pp. 126-323)

Module 10: Statistical Analysis

Technical Proposal and SOW

    1. Determination of the Significance of the Markers Discovered in the Evidentiary Material of Amerithrax Investigation (pp. 3-21)

Final Report

    2. Statistical Report of Amerithrax Data (Sept 30, 2008) (pp. 23-135)

Module 11: Cross Contamination

Reports

    1. "Risk Assessment of Anthrax Threat Letters," Defense R&D Canada, Technical Report DRES-TR-2001-048, September 2001 (pp. 3-36)

2. "Forensic Application of Microbiological Culture Analysis to Identify Mail Intentionally Contaminated with *Bacillus anthracis* Spores," D J. Beecher, Applied Environmental Biology, Aug 2006, p 5304-5310 (pp. 37-43)

Module 12: Declassified Reports

Technical Review Panel Meetings (NAS-1)

1. November 14, 2001 Technical Review Panel Meeting (pp. 3-13)

Agar and Heme Analysis (NAS-1): Agar Assay Validation - BMI

2. Statement of Work B-Task-04 Chemical Process Troubleshooting (pp. 14-20)
3. SPOT Report on B-Task-04 Technical Progress Summary (pp. 21-27)

Chemical and Physical Characteristics (NAS-2) (see Batch 2, Module 13)

Module 13: Chemical and Physical Properties

Reports

1. Determination of Concentration of Culturable Bacteria in Sample 02.57.03(Daschle) Oct 17-18, 2001 (pp. 3-7)
2. Preliminary SPOT Report on Particle Size Analysis Oct 18, 2001 (pp. 8-22)
3. SEM Images Sample A (Daschle) Oct 19, 2001 (pp. 23-28)
4. Summary of Microbiological Analyses Oct 19, 2001 (pp. 29-31)
5. Preliminary SPOT Report on Sample Analyses Oct 22, 2001 (pp. 32-57)
6. Preparation Steps and Associated Equipment Oct 25, 2001 (pp. 58-61)
7. Analysis of Silicon and Silica in Powder Samples November 21, 2001 (pp. 62-74)
8. Analysis of Silicon and Silica in Powder Samples SEM/EDS Analysis Nov 26,2001 (pp. 75-88)
9. The Analysis of Surrogate Dry Powder Bacillus Spore Product December 28, 2001 (pp. 89-98)
10. Analysis of SPS02.266.02C (Leahy) Feb 12, 2002 (pp. 99-144)
11. Summary of Sample Analysis (SPS02.266.02C) 28 Feb 02 (pp. 145-178)

Module 14: ASM Bio Defense Meeting Presentations

CBI, Mr. Thomas A. Reynolds

> 1. The Science Behind the Amerithrax Investigation: Morph A1 and A3 Assays (pp. 3-11)

MRI

> 2. Morphotype D Assay Development and Validation (pp. 13-33)

NAU, Dr. Paul Keim

> 3. The Ames Strain: Frequency, Distribution, and Forensic Analysis (pp. 35-51)

SNL, Dr. Joseph R. Michael

> 4. Elemental Microanalysis of *Bacillus anthracis* Spores from the Amerithrax Case (pp. 53-71)

TIGR, Dr. Jacques Ravel

> 5. The Genomics Behind the Amerithrax Investigation (pp. 73-106)

BATCH 3[3]

1. Amerithrax Science Update – 04-26/2002 – 11/25/2005 (271 pages)
2. EC dated 2/8/2005 – Technical Review Panel Meeting Agenda, Anthrax Review Panel, Amerithrax Panel Summary 11/14/2001; 12 page PowerPoint (17 pages)
3. EC dated 12/21/2001 – Meeting of Analytical Chemistry Expert (279A-WF-222936-LAB serial 37 and 1A 579) (7 pages)
4. EC dated 2/11/2002 – Proposed Lab Analysis and R&D Strategy Analytical Flow Chart (279A-WF-22936-LAB serial 65) (5 pages)
5. EC dates 11/14/2005 – Case Agent Meeting (attached WFO Forensic & Investigation update meeting 11/05/2001) (279A-WF-22936-LAB serial 1308 and 1A 6553) (30 pages)
6. EC dated 11/14/2005 – Scientific Review Panel Meeting June 11-22-2002 (279A-WF-22936-LAB serial 1310 and 1A 6554) (22 pages)

---

[3] The FBI did not assign page numbers to this batch of materials.

7. EC dated 12/21/2001 – External Expert Review of Analytical Plan (279A-WF-22936-LAB serial 25 and 1A 533) (7 pages)
8. U.S. Army Medical Research Institute of Infectious Diseases Reference Material Receipt Record (1 page)
9. 279A-WF-222936-BATTELLE Serial #91[4]
10. 279A-WF-222936-SC118 Serial #3 (8 pages)
11. 279A-WF-222936-USAMRIID/BEI Serial #19 (7 pages)
12. 279A-WF22936-USAMRIID Serial #1418 (2 pages)
13. Dr. Ivins USAMRIID Laboratory Notebook #4010 (30 pages)
14. Incoming Shipment Records for the 8 Positive FBI Repository Samples (origin is FBIR Database) (19 pages)
15A. United States District Court Search Warrant Application and Affidavit Case#07-524-M-01 (33 pages)
15B. United States District Court Search Warrant Application and Affidavit Case#07-525-M-01 (27 pages)
15C. United States District Court Search Warrant Application and Affidavit Case#07-526-M-01 (28 pages)
15D. United States District Court Search Warrant Application and Affidavit Case#07-527-M-01 (28 pages)
15E. United States District Court Search Warrant Application and Affidavit Case#07-528-M-01 (28 pages)
15F. United States District Court Search Warrant Application and Affidavit Case#07-529-M-01 (39 pages)
16. FBI Repository Shipment Records[5] (32 pages)

SUPPLEMENTAL DOCUMENTS[6]

1. AFIP Materials related to USAMRIID Specimens October 2001 (41 pages)
2. Preparing and Shipping TSA Slants for *B. Anthracis* Ames (1 page)
3. FBI WFO [Washington Field Office] Report on Samples from an Overseas Site Identified by Intelligence[7] (18 pages)

---

[4] While this document was listed in the Batch 3 table of contents provided to the committee by the FBI, this document was not provided to the committee because of its security classification.

[5] The FBI did not assign a name to this document. The descriptor was assigned by the committee.

[6] The FBI did not assign page numbers to supplemental documents provided to the committee.

[7] The FBI did not assign a name to this document. The descriptor was assigned by the committee.

Appendix A

Radiocarbon Dating

The technique of radiocarbon dating was pioneered by Libby in the early 1950s. This technique is based on the half life of ^{14}C of approximately 5,700 years and the fact that the radioisotope is made through cosmic ray interactions in the atmosphere by the reaction of energetic neutrons on ^{14}N to produce ^{14}C and a proton. The ^{14}C mixes in the atmosphere and becomes incorporated into CO_2. All living things have ratios of $^{14}C/^{12}C$ that reflect the value in the atmosphere at the time and place in which they are living. Once a plant or animal dies, it stops incorporating modern CO_2 and the ^{14}C slowly decays away. The standard is to reference everything to the ratio of $^{14}C/^{12}C$ in the atmosphere in 1950. The units are in the change relative to this value as $\Delta^{14}C$ in units or ‰ (parts per thousand.) All positive ratios are from after 1950 and all negative ratios correspond to before 1950, due to the loss of ^{14}C because of its radioactive decay. The complication is in determining what the $^{14}C/^{12}C$ was at the time the animal or plant was living. The historical ratio is influenced by the cosmic ray flux in a given geographical region (it is stronger near the earth's poles and sensitive to the earth's magnetic field) and the solar activity that is a dominant source of the earth's cosmic rays. Tree rings and corals have been used as calibrations to relate the $\Delta^{14}C$ to actual calendar ages.

Modern humans have modified the purely cosmogenic ^{14}C content of the atmosphere in two ways. Fossil fuel burning has introduced into the atmosphere old (dead) carbon that is so old that it has essentially no ^{14}C left. Therefore, the "modern" $^{14}C/^{12}C$ ratio was falling until the mid-1950s, when atmospheric nuclear testing created what is now called the "bomb spike." (see Figure A-1) The testing raised the ^{14}C content of the atmosphere to almost twice the pre-bomb value in the Northern hemisphere at its peak around 1965. With the cessation of atmospheric testing, the exchange of the atmosphere with the ocean has gradually reduced the levels to values of $\Delta^{14}C$ less than 100‰, down from the peak of >800 ‰. This behavior is clearly seen in the figure below, which shows a modern $\Delta^{14}C$ curve using data from Levin and Kromer (2004). This

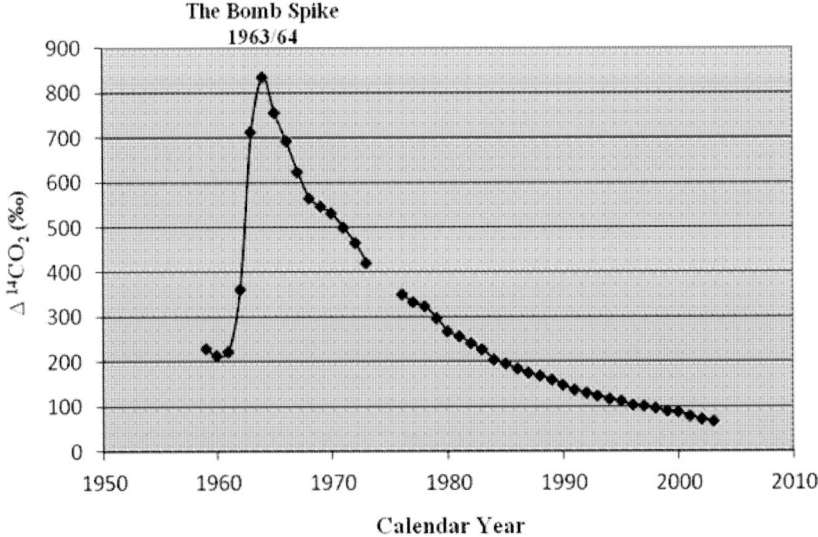

FIGURE A-1 Atmospheric CO_2 (Northern Hemisphere).
Change in values of ^{14}C for atmospheric CO_2 since 1959. The rise from values near zero results from atmospheric testing of nuclear weapons. The decline from peak values reached in late 1963 results from the exchange of CO_2 between the atmosphere and ocean. As a result of this exchange, levels of ^{14}C in the ocean have slowly risen. The gap in the curve between 1973 and 1976 is due to a lack of atmospheric data for 1974 and 1975.
SOURCE: Courtesy of Alice Mignerey.

rapid rise and fall enables the dating of modern (younger than 1950) samples to within a few years in some cases.

Radiocarbon dating has advanced tremendously with the advent of the technique of accelerator mass spectrometry (AMS) to identify individual ^{14}C atoms. This is a direct counting method and does not rely on detecting the radiation that is emitted when the ^{14}C atoms decay. This improvement has led to the capability to radiocarbon date samples of less than 1 mg in mass. This new technique enabled the anthrax samples to be analyzed.

Appendix B

The Forensics Potential of Stable Isotope Analysis

Typically, elements in nature behave the same chemically and biologically, independent of their isotopic identity. However, there are differences in a number of physical, chemical, and biological processes that produce small variations in the ratios of minor isotopes of elements to their major components. These differences are small, but with good analytical instrumentation they can be detected. One example is the process of evaporation and condensation. Water that has been evaporated from a large source, such as the ocean, tends to be deficient in the heavier isotopes of hydrogen and oxygen, relative to the original source. In the reverse process of condensation the heavier isotopes will condense more readily, yielding rain that is enriched in the heavier isotopes. This continual cycle of evaporation and condensation makes the ratio of $^2H/^1H$ and $^{18}O/^{16}O$ lower for water sources that are farther removed from the oceans (the major ultimate-water source). The results are also a function of temperature (summer versus winter precipitation) and altitude. An extensive sampling of river waters in the United States was performed to provide a baseline for further studies (Kendall and Coplen, 2001). River water can come from a number of sources—groundwater as well as surface runoff—but even with this complexity there are some clear geographical trends in the resultant maps of the deviation of the ^2H and ^{18}O composition of the water when compared to mean ocean water, expressed as δ^2H and $\delta^{18}O$ in ‰ (parts per thousand). Negative values mean that the water is depleted in the heavier isotopes relative to ocean water. Since this process affects both the $^2H/^1H$ and $^{18}O/^{16}O$ ratios, a plot of δ^2H versus $\delta^{18}O$ gives a roughly straight line called the meteoric water line (MWL). Samples that fall below this line are usually from arid regions with low humidities. While different regions have different local linear relationships, the overall trend can be used to infer an average value of $\delta^{18}O$ from a given δ^2H value.

The isotopes of carbon are influenced by the biological processes that are used in the synthesis of the organic compounds by an organism. There are very different ratios of $^{13}C/^{12}C$ for the C3 and C4 photosynthetic pathways. A C3 plant will typically exhibit a $\delta^{13}C$ around –25‰, when referenced against a standard limestone, while a C4 plant will give $\delta^{13}C$ values around $-9‰$. A $\delta^{13}C$ of about $-16‰$ might represent either a mixture of the two sources or a plant that has grown in water (algae or a hydroponically grown plant). Animals will acquire the $\delta^{13}C$ signature of their food sources. In the case of bacteria cultured in the laboratory, the signature of the growth medium will be reflected in that of the spores produced. Tests of this expectation for liquid growth medium bear this out (Kreuzer-Martin and Jarman, 2007).

In the case of δ^2H and $\delta^{18}O$, cultured microorganisms record the isotopic signature of the water in the culture medium as well the nutrients in the medium. Since the anthrax attacks of 2001 there have been extensive studies testing these relationships using the nonpathogenic *Bacillus subtilis*. On average about 70 percent of the oxygen atoms in spores produced come from the water, while only about 30 percent of the hydrogen atoms come from the water (Kreuzer-Martin et al., 2003, 2005) Results show a strong positive correlation between the δ^2H and $\delta^{18}O$ content of the culture water and resultant spores for liquid cultures (Kreuzer-Martin and Jarman, 2007; Kreuzer-Martin et al., 2003). Samples grown on an agar medium are also subjected to isotopic fractionation from evaporation of water from the medium. It was found that the influence of evaporation was small for δ^2H, but significant for $\delta^{18}O$ (Kreuzer-Martin et al., 2005). Exchange of water vapor with the ambient atmosphere could also be important if the agar was prepared in one location and used in another, or anytime there is a different isotopic signature of the water used and that of the surrounding water vapor, as in the case of studies at Lawrence Livermore National Laboratory, where the source of the tap water is largely the Sierra Mountains and the ambient atmosphere can have a significant marine input (Kreuzer-Martin et al., 2005)

The forensic value of isotopic measurements on bacteria cultures depends on the individual circumstances. It has the potential to rule out certain combinations of water and growth media as well as to provide a distinguishing marker for discrimination between production batches of spores.

Appendix C

Committee Evaluation of *Statistical Analysis Report*

The *Statistical Analysis Report* (B2M10) was submitted in response to a contract with the FBI, to analyze the results of the assays on the 1,070 FBI Repository (FBIR) samples and determine whether the results of the assays could be related to those obtained on the evidentiary material. If such a relationship could be identified, a secondary issue was the development of a measure of its "statistical strength." Two of the attack letters assayed positive for the four mutations A1, A3, D, and E. Results on 1,059 of the 1,070 samples were tabulated in the *Statistical Analysis Report*. Eight samples tested positive for all four mutations; seven of these eight samples came from one institution (USAMRIID) and the remaining sample came from a different institution (Battelle Memorial Institute [BMI]). A table of documented transfers of samples from one institution to another showed a transfer of sample material from the first institution (USAMRIID) to the second institution (BMI). This Appendix discusses the validity of the inferences and calculations in the *Statistical Analysis Report* submitted to the FBI.

As noted in Chapter 6, the statistical analyses used in the report (e.g., 95 percent confidence interval for the proportion of samples with four mutations, chi-squared tests of independence) require two key assumptions to be valid:

1. **Representativeness**: The 1,059 samples are assumed to be a representative and random collection of samples from some well-defined population of samples.
2. **Independence**: The 1,059 samples are assumed to be independent of one another (i.e., have no connection with each other, beyond that they all come from the same population).

The *Statistical Analysis Report* acknowledges that neither assumption can be validated from these data. The committee agrees with this assessment. As a consequence, many of the statistical methods applied to these data cannot be validated. The consequences of the violation of these assumptions and their impacts are listed below.

1. *FBIR is not a representative and random collection of samples from a well-defined population of* B. anthracis *samples.*

   The 1,059 samples do not appear to satisfy assumption 1. They were obtained in response to a request from the FBI. No information is available on samples in the population that were not submitted. In fact, the "target population" seems not to have been defined. It could be the population of all unique preparations of *B. anthracis* Ames in the United States, or in the world, or from selected institutions. The absence of a definition of "well-defined population" makes it difficult to assess representativeness of the collection. The elimination of samples that had "inconclusive" results on assays also appears to be nonrandom, as some institutions had many more "inconclusive" assays than others.

2. *The 1,059 samples in the FBIR are not independent.*

   The FBI submitted to the committee a table of known transfers of samples between institutions. Hence, the second assumption is violated. Thus, the results of the chi-squared tests for independence of the mutations that are calculated in the report are not meaningful. Further, the confidence interval for the proportion 8/947 is not appropriate. The correct denominator for this proportion is likely not 947. A more accurate numerator and denominator might refer to the number of known independent preparations rather than the number of samples, but such information may not be possible to obtain.

3. *Violation of assumptions renders invalid the inferences from the statistical analyses.*

   Because the FBIR is not a representative and random collection of independent samples, the results on the assays from the repository may be biased. Virtually all statistical procedures assume that the units on which measurements are made comprise a random, representative collection from the target population. (The effects of biased sampling on inferences have been well documented; see, e.g., Freedman et al., 2007). Without an appropriate model that characterizes the nonrepresentativeness and the degree of dependence among the samples, it is not possible to calculate a meaningful measure of "statistical significance" in the results.

4. Results on 112 samples beyond the 947 samples

The *Statistical Analysis Report* eliminated from most of its tables the results of the assays on 112 samples that showed "inconclusive" for A1, A3, MRI-D, or E. Twenty-one of these 112 sample that were eliminated from the statistical analysis assayed positive for 1, 2, or 3 mutations. Table C-1 lists these samples. (Five samples—05-022, 49-014, 53-014, 53-068, 54-008—are listed twice because they were reported as "inconclusive" or "variant" on two assays.)

TABLE C-1 Samples with Positive and "Inconclusive" or "Variant" Assays

FBIR Number	A1	A3	MRI-D	IITRI-D	E	+Mutations
039-010	inc	+	-	-	-	A3
044-034	var	-	-	+	-	IITRI-D
049-014	inc	var	+	+	-	MRI-D, IITRI-D
053-004	var	+	-	-	+	A3, E
053-010	var	+	+	+	+	A3, MRI-D, IITRI-D, E
053-014	var	inc	-	-	+	E
053-068	inc	inc	+	-	-	MRI-D
054-008	inc	+	+	inc	+	A3, MRI-D, E
061-030	inc	-	+	+	-	MRI-D, IITRI-D
066-015	inc	inc	+	+	-	MRI-D, IITRI-D
005-022	+	var	-	inc	+	A1, E
017-006	-	var	+	+	-	MRI-D, IITRI-D
049-014	inc	var	+	+	-	MRI-D, IITRI-D
049-018	-	var	+	+	-	MRI-D, ITRI-D
053-014	var	inc	-	-	+	E
053-068	inc	inc	+	-	-	MRI-D
054-066	+	inc	+	+	+	A1, MRI-D, IITRI-D, E
054-068	-	var	+	-	+	MRI-D, E
005-020	-	-	+	inc	-	MRI-D
005-022	+	var	-	inc	+	A1, E
043-016	-	-	+	inc	-	MRI-D
044-020	-	+	-	inc	+	A3, E
052-026	+	+	+	inc	-	A1, A3, MRI-D
054-008	inc	+	+	inc	+	A3, MRI-D, E
054-022	-	-	+	inc	-	MRI-D
057-036	-	-	+	inc	-	MRI-D

inc = inconclusive
IITRI = Illinois Institute for Technology Research Institute
MRI = Midwest Research Institute
var = variant

In addition to the two 3-positive samples (+++) among the 947 samples, the four samples below also tested positive for 3 mutations (ordered by FBIR number):

052-026	+	+	+	inc	-	A1, A3, MRI-D
053-010	var	+	+	+	+	A3, MRI-D, IITRI-D, E
054-008	inc	+	+	inc	+	A3, MRI-D, E
054-066	+	Inc	+	+	+	A1, MRI-D, IITRI-D, E

The following four samples revealed positive assays for 2 of the 4 mutations, in addition to the 11 samples noted among the 947 samples (ordered by FBIR number):

005-022	+	var	-	inc	+	A1, E
044-020	-	+	-	inc	+	A3, E
053-004	var	+	-	-	+	A3, E
054-068	-	var	+	-	+	MRI-D, E

DILUTION EXPERIMENTS

Dilution experiments were conducted to assess the sensitivity of the assays to various concentrations. Thirty samples were prepared from RMR-1029 at dilution 10.0. As with the other samples, some of the assays were "inconclusive." Genotype E tested positive in all 30 samples; all 4 mutations tested positive for 16 samples. But in the remaining 14 samples, assays for one or more of the genotypes were negative. In fact, one sample tested negative for A1, A3, and D; it was positive for only E. Five samples were positive for two mutations only (A3 and E), and eight samples were positive for only three of the four mutations (7 for A3, D, E; 1 for A1, A3, E). Thus, 6 of the 30 replicate samples (20 percent) tested positive for only 1 or 2 of the mutations. Given that 50 of the 947 FBIR samples showed only 1 positive, and 11 of the 947 showed only 2 positives, this variation indicates that some of the samples may have harbored mutations that went undetected. Absent any repeat testing of these samples, however, it is difficult to know how such false negatives might have affected the inferences.

Additional experiments were conducted on RMR-1029 and another sample, "SPS.266 Tube#5," at 10 dilutions levels (10.1, . . ., 10.10). The results of the three replicates at each dilution level, for each of the five genotypes, for samples from both RMR-1029 and SPS.266 Tube#5 were reported in Chapter 6. Variability in the results on replicates, even from the same sample at the same dilution level, demonstrates the value, and need for, replicate testing. For example, the results on the three replicates from RMR-1029 at dilution 10.1, ordered as A1, A3, MRI-D, IITRI-D, E, were: (- + + + +), (- + + + -), (+ + + + -). Clearly, dilution affects the assay result: the greater the dilution, the more likely the assay is negative. Moreover, it is perhaps unexpected that greater dilutions sometimes give positive results when not all replicates at lesser dilutions did so.

CONCORDANCE OF TESTS FROM IITRI-D AND MRI-D

The FBI retained both the Illinois Institute for Technology Research Institute (IITRI) and Midwest Research Institute (MRI) to conduct the D assays. Because the assays on the 1,059 samples can be considered to be independent between IITRI and MRI, the *Statistical Analysis Report* (Table 3, p. 7, as presented below) tabulates the results of the D assays from the two facilities:

MRI-D	IITRI-D					
	Inconclusive	Negative	No growth	Pending	Positive	Total
Inconclusive	0	22	12	0	0	34
Negative	17	909	1	1	12	940
Negative-u	1	20	0	0	0	21
Positive	6	12	0	0	46	64
TOTAL	24	963	13	1	58	1,059

The *Statistical Analysis Report* combined the "negative-u" results with the "negative" results, and eliminated the 12 samples that showed "no-growth" by IITRI-D and "inconclusive" by MRI-D as well as the one "pending" sample, to yield the following table:

MRI-D	IITRI-D			
	Inconclusive	Negative	Positive	Total
Inconclusive	0	22	0	22
Negative	18	929	12	959
Positive	6	12	46	64
TOTAL	24	963	58	1,045

Eliminating the 14 "no-growth" and "pending" samples, the concordance rate is $975/1045 = 0.933$, with a 95 percent confidence interval (0.916, 0.947). Thus, the agreement between the facilities is unlikely to be lower than 91.6 percent and likely does not exceed 94.7 percent. Of greater interest, however, are the 12 samples that were positive by IITRI-D but negative for MRI-D, the 12 samples that were negative by IITRI-D but positive by MRI-D, and the six samples that were positive by MRI-D but inconclusive by IITRI-D. While concordance is informative, these 30 samples with discordant results might provide increased information about the samples and the assay process. On the other hand, we also know from the repeated assays of the dilution series that some discordance also arises owing to variation even when using the same assay procedure.

In any case, because genotype D is the only one of the four genotypes that was subjected to independent testing by a second organization one cannot say whether the results on the other genotypes might have been different if they also had been subjected to independent testing.

"SIGNIFICANCE" OF SEVEN (++++) SAMPLES FROM INSTITUTION F

The *Statistical Analysis Report* notes in its conclusions:

> "In summation, though the random chance of occurrence of the sample type (++++) is 8 out of 947 (i.e., 0.84%) with exact 95% confidence interval of 0.0037 to 0.0166 (I.e., from 1 in 270 to 1 in 60), this sample type has been found in only two institutions thus far sampled (USAMRIID and BMI), and its occurrence in BMI is explained by a recent sample transfer from USAM to BMI, since there is no documented record of sample transfers in the other direction." (p. 2)

As noted in Chapter 6, 598 of the 947 samples (63 percent) came from Institution F. (Twelve of the institutions submitted 6 or fewer samples; 4 institutions submitted 15-31 samples, and 4 institutions submitted 49-74 samples.) Therefore, one would not be surprised to find more "mutation-positive" samples from Institution F than, say, from Institution B (which contributed only one sample). One might naturally ask: How unusual is the occurrence of seven "4-mutation" samples—or even all eight—from Institution F? Given that Institution F contributed almost 2/3 of the 947 samples, how many of the 4-positive (++++) samples would Institution F receive if the 4-mutation samples were distributed completely at random?

The answer to this question is given by the probabilities of observing 0 or 1 or 2 or ... or 8 of the eight (++++) samples from Institution F, given that Institution F submitted 598 of the 947 samples that yielded definitive results on the A1, A3, MRI-D, and E assays. These probabilities (from the hypergeometric probability distribution) (Johnson et al., 2005) are shown in Table C-2.

TABLE C-2 Probabilities of k 4-Mutation Samples in Institution F

k =	0	1	2	3	4	5	6	7	8
Probability	0.0003	0.0045	0.0276	0.0955	0.2058	0.2826	0.2415	0.1174	0.0248

This table shows that the chance of Institution F having ended up with seven or eight of the eight (++++) samples is (0.1174 + 0.0248) = 0.1422, or about 1/7. Therefore, while the observed data showing that seven of the eight (++++) samples appeared in Institution F is not completely typical, it also could hardly be considered extreme.

Appendix D

Biographical Information of Committee and Staff

Alice P. Gast (*Chair*) (NAE) became Lehigh University's 13th president on August 1, 2006. Previously she was the Robert T. Haslam Professor of Chemical Engineering, Vice President for Research, and Associate Provost at Massachusetts Institute of Technology. Prior to moving to MIT in 2001, she spent 16 years as a professor of chemical engineering at Stanford University and at the Stanford Synchrotron Radiation Laboratory. In her research she studies surface and interfacial phenomena, in particular the behavior of complex fluids. Some of her areas of research include colloidal aggregation and ordering, protein lipid interactions, and enzyme reactions at surfaces. In 1997 Dr. Gast coauthored the sixth edition of *Physical Chemistry of Surfaces* with Arthur Adamson. Professor Gast received her B.S. in Chemical Engineering from the University of Southern California. After earning her Ph.D. in chemical engineering from Princeton University, Gast spent a postdoctoral year on a NATO fellowship at the École Supérieure de Physique et de Chimie Industrielles in Paris. She returned there for a sabbatical as a Guggenheim Fellow. She was a 1999 Alexander von Humboldt Fellow at the Technical University in Garching, Germany. She received the National Academy of Sciences Award for Initiative in Research, and the Colburn Award of the American Institute of Chemical Engineers. She was elected to the National Academy of Engineering in 2001 and to the American Academy of Arts and Sciences in 2002. Professor Gast has served on numerous advisory committees and boards, including the Board of the American Association for the Advancement of Science (AAAS) and the National Space Biomedical Research Institute Board of Directors. She is a member of the American Chemical Society, the American Institute of Chemical Engineers, and the American Physical Society.

David A. Relman (*Vice Chair*) is the Thomas C. and Joan M. Merigan Professor in the Departments of Medicine and of Microbiology and Immunology at Stanford University, and Chief of Infectious Diseases at the VA Palo Alto Health Care System in Palo Alto, California. He received an S.B. (Biology) from MIT

(1977) and M.D. (magna cum laude) from Harvard Medical School (1982), completed his clinical training in internal medicine and infectious diseases at Massachusetts General Hospital, served as a postdoctoral fellow in microbiology at Stanford University, and joined the faculty at Stanford in 1994. Dr. Relman's current research focus is the human indigenous microbiota (microbiome), and in particular the nature and mechanisms of variation in patterns of microbial diversity in the human body as a function of time (microbial succession) and space (biogeography within the host landscape) and in response to perturbation, e.g., antibiotics (community robustness and resilience). One of the goals of this work is to define the role of the human microbiome in health and disease. This research integrates theory and methods from ecology, population biology, environmental microbiology, genomics, and clinical medicine. During the past few decades, his research directions have also included pathogen discovery and the development of new strategies for identifying previously unrecognized microbial agents of disease. This work helped to spearhead the application of molecular methods to the diagnosis of infectious diseases in the 1990s. His research has emphasized the use of genomic approaches for exploring host-microbe relationships. Past scientific achievements include the description of a novel approach for identifying previously unknown pathogens, the identification of a number of new human microbial pathogens, including the agent of Whipple's disease, and some of the most extensive and revealing analyses to date of the human indigenous microbial ecosystem. Dr. Relman advises the U.S. government as well as nongovernmental organizations in matters pertaining to microbiology, emerging infectious diseases, and biosecurity. He currently serves as Chair of the Institute of Medicine's Forum on Microbial Threats (National Academy of Sciences), a member of the National Science Advisory Board for Biosecurity (NSABB), and a member of the Physical and Life Sciences Directorate Review Committee for Lawrence Livermore National Laboratory, and advises several U.S. government departments and agencies on matters related to pathogen diversity, the future life sciences landscape, and the nature of present and future biological threats. He has served as Chair of the Board of Scientific Counselors of the National Institute of Dental and Craniofacial Research (National Institutes of Health [NIH]) and member of the Board of Directors, Infectious Diseases Society of America (IDSA). Dr. Relman cochaired a three-year National Research Council study that produced a widely cited report entitled *Globalization, Biosecurity, and the Future of the Life Sciences* (2006). He is a Fellow of the American Academy of Microbiology and a member of the Association of American Physicians. Dr. Relman received the Squibb Award from the IDSA in 2001 and was the recipient of both the NIH Director's Pioneer Award and the Distinguished Clinical Scientist Award from the Doris Duke Charitable Foundation in 2006.

Arturo Casadevall is the Leo and Julia Forchheimer Professor of Microbiology and Immunology and Chair of the Department of Microbiology and Immunology at the Albert Einstein College of Medicine. He is also a Professor in the Department of Medicine. He received his B.A. from Queens College, CUNY, and M.S., M.D., and Ph.D. degrees from New York University. His laboratory has a multidisciplinary research program spanning several areas of basic immunology and microbiology to address general questions in these areas, resulting in over 460 publications. His laboratory studies are focused on two microbes: the fungus *Cryptococcus neoformans*, a ubiquitous environmental microbe that is a frequent cause of disease in immunocompromised individuals, and *Bacillus anthracis*, which is a major agent of biological warfare. He is a fellow of the American Academy of Microbiology and has been elected to the American Society for Clinical Investigation, to the American Association of Physicians, and as a fellow of AAAS. Dr. Casadevall has served on numerous NIH advisory committees including study sections, strategic planning for the National Institute of Allergy and Infectious Diseases (NIAID), and the blue ribbon panel on response to bioterrorism. He currently cochairs the NIAID Board of Scientific Counselors and is a member of the NSABB. He is the founding editor of the first American Society of Microbiology (ASM) general journal, *mBio*, serves on the editorial boards of several journals, and has been the recipient of numerous awards, most recently the Solomon A. Berson Medical Alumni Achievement Award in Basic Science-NYU School of Medicine 2005, IDSA Kass Lecturer in 2008, and the ASM William Hinton Award for mentoring scientists from underrepresented groups.

Nancy D. Connell is professor of medicine at the University of Medicine and Dentistry of New Jersey (UMDNJ)-New Jersey Medical School. She is also director of the UMDNJ Center for BioDefense, which was established in 1999 for research into the detection and diagnosis of biological warfare agents and biodefense preparedness. Dr. Connell also is director of the Biosafety Level 3 Facility of UMDNJ's Center for the Study of Emerging and Re-emerging Pathogens and chairs the university's Institutional Biosafety Committee. She is past chair of NIH's Center for Scientific Review Study Section HIBP (Host Interactions with Bacterial Pathogens), which reviews bacterial pathogenesis submissions to NIAID. She is current chair of the F13 infectious diseases and microbiology fellowship panel. Dr. Connell's involvement in biological weapons control began in 1984, when she was chair of the Committee on the Military Use of Biological Research, a subcommittee of the Council for Responsible Genetics, based in Cambridge, Massachusetts. She has worked with several international programs on dual use research issues and served on various NRC committees with her expertise in select agent microbiology, dual use, and biocontainment. Dr. Connell received her Ph.D. in microbial genetics from Harvard University. Her major research focus is the interaction between *Mycobacterium tuberculosis* and the macrophage.

Thomas V. Inglesby is CEO and Director of the Center for Biosecurity of the University of Pittsburgh Medical Center and Associate Professor of Medicine and Public Health at the University of Pittsburgh Schools of Medicine and Public Health. He is an infectious disease physician by training. He is Coeditor in Chief of the peer-reviewed journal *Biosecurity and Bioterrorism: Biodefense Strategy, Practice, and Science* and has authored a number of widely cited publications on anthrax, smallpox, plague, and biosecurity issues related to medicine and hospital preparedness, public health, science, pandemic planning, and national security. He is a principal editor of the *Journal of the American Medical Association* book entitled *Bioterrorism: Guidelines for Medical and Public Health Management*. Dr. Inglesby was a principal designer, author, and controller of the widely recognized Atlantic Storm exercise of 2005 and of the Dark Winter smallpox exercise of 2001. He has served in advisory and consultative capacities for government, scientific organizations, and academia on issues related to biosecurity, providing briefings for officials in the administration and for congressional members and staff; serving on a task force of the Defense Science Board of the Department of Defense and a committee of the US National Research Council; and participating in an advisory capacity to the Centers for Disease Control and Prevention, NIH, the Department of Health and Human Services, the Department of Homeland Security, the Defense Advanced Research Projects Agency (DARPA), and the Defense Intelligence Agency (DIA). Prior to helping establish the Center for Biosecurity in 2003, Dr. Inglesby was one of the founding members of the Johns Hopkins Center for Civilian Biodefense Strategies, where he served as Deputy Director from 2001 to 2003. He was also a faculty member of the Johns Hopkins School of Medicine from 1999 to 2003. Dr. Inglesby is Board-certified in Infectious Diseases. He received a B.A. in 1988 from Georgetown University and an M.D. from the Columbia University College of Physicians and Surgeons in 1992. He completed his internal medicine residency and Infectious Diseases Fellowship training at the Johns Hopkins School of Medicine, and served as Assistant Chief of Service in the Johns Hopkins Department of Medicine in 1996 and 1997.

Murray V. Johnston is Professor of Chemistry in the Department of Chemistry and Biochemistry at the University of Delaware. He received a B.S. in chemistry from Bucknell University and Ph.D. in analytical chemistry from the University of Wisconsin, Madison. He is the recipient of a Center for Advanced Study fellowship in 1999, the Outstanding Scholar Award in the College of Arts and Sciences in 2001, the Delaware Section Award of the American Chemical Society in 2003, and the Benjamin Y.H. Liu Award from the American Association for Aerosol Research in 2008. In 2007, he served on the National Research Council panel on Testing and Evaluation of Biological Standoff Detection Systems. Dr. Johnston's research includes applications of mass spectrometry to a wide array of materials, from airborne particles to biological and polymeric macromolecules. He has used real-time single-particle mass spectrometry to study microchemical reactions within particles, heterogeneous reactions between gas phase and particulate-phase species, and ambient particles at various urban sites. His current work emphasizes the use of photoionization aerosol mass spectrometry to characterize organic components of combustion and ambient aerosols, nano aerosol mass spectrometry to characterize individual nanoparticles and macromolecules smaller than about 30 nm, and conventional mass spectrometry to characterize oligomeric compounds in secondary organic aerosols. Dr. Johnston is a member of the editorial board of the journal *Analytical Chemistry* and the Board of Directors of the American Association for Aerosol Research. He has served as an ad hoc member of several NIH review panels associated with biological and environmental mass spectrometry.

Karen Kafadar is James H. Rudy Professor of Statistics and Physics at Indiana University. She received her B.S. and M.S. degrees from Stanford and her Ph.D. in Statistics from Princeton under John Tukey. Her research focuses on exploratory data analysis, robust methods, characterization of uncertainty in quantitative studies, and analysis of experimental data in the physical, chemical, biological, and engineering sciences. Prior to Indiana University, she was Professor and Chancellor's Scholar in the Departments of Mathematical Sciences and Preventive Medicine and Biometrics at the University of Colorado-Denver; Fellow at the National Cancer Institute (cancer screening section); and Mathematical Statistician at Hewlett Packard Company (R&D laboratory for RF/microwave test equipment) and at the National Institute of Standards and Technology (where she continues as Guest Faculty Visitor on problems of measurement accuracy, experimental design, and data analysis). Previous engagements include consultancies in industry and government as well as visiting appointments at University of Bath, Virginia Tech, and Iowa State University. She has served on previous NRC committees and chaired the Committee on Applied and Theoretical Statistics. She also serves on the editorial boards for several professional journals as Editor or Associate Editor and on the governing boards for the American Statistical Association, the Institute of Mathematical Statistics, and the International Statistical Institute. She is an Elected Fellow of the American Statistical Association and the International Statistical Institute, has authored over 90 journal articles and book chapters, and has advised numerous M.S. and Ph.D. students.

Richard E. Lenski is the John A. Hannah Distinguished Professor of Microbial Ecology at Michigan State University. His research explores the genetic mechanisms and ecological processes that underlie evolution. While most evolutionary research uses the comparative method, Lenski pursues an experimental approach using bacteria. In an experiment started in 1988, Lenski and his team have watched 12 populations of *E. coli* evolve in the lab for more than 50,000 generations to investigate the phenotypic and genetic dynamics of adaptation and diversification. Lenski and his students have performed other experiments with microbes on the dynamics of host-parasite interactions, the evolution of mutation rates, and even social interactions. Lenski also collaborates with an interdisciplinary team on experiments using digital organisms—computer programs that replicate, mutate, compete, and evolve—to investigate the evolution of complexity. Professor Lenski received a MacArthur Foundation Fellowship in 1996 and was elected to the National Academy of Sciences in 2006.

Richard M. Losick is the Maria Moors Cabot Professor of Biology, a Harvard College Professor, and a Howard Hughes Medical Institute Professor in the Faculty of Arts and Sciences at Harvard University. He received his A.B. in Chemistry at Princeton University and his Ph.D. in Biochemistry at the Massachusetts Institute of Technology. Upon completion of his graduate work, Professor Losick was named a Junior Fellow of the Harvard Society of Fellows when he began his studies on RNA polymerase and the regulation of gene transcription in bacteria. Professor Losick is a past Chairman of the Departments of Cellular and Developmental Biology and Molecular and Cellular Biology at Harvard University. He received the Camille and Henry Dreyfuss Teacher-Scholar Award and is a member of the National Academy of Sciences, a Fellow of the American Academy of Arts and Sciences, a member of the American Philosophical Society, a Fellow of the American Association for the Advancement of Science, a Fellow of the American Academy of Microbiology, and a former Visiting Scholar of the Phi Beta Kappa Society. He is the 2007 recipient of the Selman A. Waksman Award of the National Academy of Sciences and a 2009 recipient of the Canada Gairdner Award.

Alice C. Mignerey is a nuclear chemist with research programs in basic nuclear science and in applications of the nuclear analytical technique of accelerator mass spectrometry (AMS) to environmental problems. Professor Mignerey's basic nuclear research is focused on understanding the behavior of nuclear matter under conditions of extreme density (pressure) and temperature. These conditions are postulated to have existed just after the Big Bang, when the protons and neutrons had not yet formed from their constituent quarks and the gluons that hold them together. This so-called quark-gluon plasma has been predicted to be accessible through heavy ion reactions at high energies. The experimental program is centered at the Brookhaven National Laboratory Relativistic Heavy Ion Collider (RHIC) accelerator where colliding beams of nuclei reach center-of-mass energies of 200 AGeV, producing conditions mimicking those of the early universe. Professor Mignerey is a member of the Phobos and PHENIX Collaborations at RHIC and the CMS Heavy Ion Group at the CERN Large Hadron Collider (LHC). The research program in AMS has concentrated on the uses of the cosmogenic nuclides, such as C-14 and Cl-36, to study groundwater and soil systems. Technique development is currently being carried out with researchers at the Naval Research Laboratory Trace Element AMS facility (TEAMS) to allow the dating of separate organic fractions in the organic C-14 carbon pool.

David L. Popham is a professor in the Department of Biological Sciences at Virginia Tech. He teaches in the areas of microbial genetics and physiology. He directs a research program in the areas of bacterial endospore structure, content, germination, and resistance properties. Dr. Popham has a Ph.D. in microbiology from the University of California-Davis. He held postdoctoral research positions at the Institut de Biologie Physico-Chimique in Paris and at the University of Connecticut Health Science Center before joining the Virginia Tech faculty in 1996. He has over 20 years of experience in research on *Bacillus subtilis* cell wall synthesis, spore formation, and spore resistance properties. More recently his research has expanded into the content, structure, and germination of spores produced by *B. anthracis*, *Clostridium difficile*, and *C. perfringens*. Dr. Popham is a member of the editorial boards of the *Journal of Bacteriology* and *Molecular Microbiology* and has served as a member of six NIH grant review panels. He has served on the Environmental Protection Agency Federal Insecticide, Fungicide, and Rodenticide Act Scientific Advisory Panel for the development of guidelines for the approval of sporicidal products.

Jed S. Rakoff has been a United States District Judge for the Southern District of New York since 1996. Prior to his appointment, he was a partner at Fried, Frank, Harris, Shriver & Jacobson LLP. From 1980 to 1990, he was a partner at Mudge, Rose, Guthrie, Alexander & Ferdon LLP. Judge Rakoff was an Assistant U.S. Attorney for the Southern District of New York from 1973 to 1980 and chief of the Business and Securities Fraud Prosecutions Unit from 1978 to 1980. Before joining the U.S. Attorney's Office, Judge Rakoff spent two years in private practice as an associate attorney at Debevoise & Plimpton LLP. He served as a law clerk to the Honorable Abraham L. Freedman, U.S. Court of Appeals, 3rd Circuit, in 1969-70. Judge Rakoff is coauthor of five books and author of more than 100 published articles, more than 300 speeches, and more than 650 judicial opinions. He has been a lecturer in law at Columbia Law School since 1988. He was a member of the Board of Managers, Swarthmore College, from 2004 to 2008. Judge Rakoff currently serves as a Trustee for the William Nelson Cromwell Foundation and a member of the Governance Board for the MacArthur Foundation Initiative on Law and Neuroscience. He is a member of the National Research Council Committee on the Development of the Third Edition of the Reference Manual on Scientific Evidence and chair of the Criminal Justice Advisory Board, Southern District of New York; the Second Circuit Bankruptcy Committee; and the Honors Committee of the New York City Bar Association. He is a Judicial Fellow at the American College of

Trial Lawyers and was chair of the Downstate New York Chapter in 1993-94. Judge Rakoff is the former director of the New York Council of Defense Lawyers and former chair of the Criminal Law Committee, New York City Bar Association. He has been a Judicial Fellow at the American Board of Criminal Lawyers since 1995. Judge Rakoff received a B.A. from Swarthmore College in 1964, an M.Phil. from Oxford University in 1966, and a J.D. from Harvard Law School in 1969. He was awarded honorary LL.D.s from Swarthmore College in 2003 and St. Francis University in 2005.

Robert C. Shaler obtained a doctoral degree in Biochemistry from the Pennsylvania State University in 1968 and then worked at the University of Pittsburgh as a professor of chemistry and at the Pittsburgh Crime Laboratory as a criminalist. His research resulted in the development of a bloodstain analysis system, the de facto standard in forensic laboratories until the early 1990s. The New York City Office of Chief Medical Examiner beckoned in 1978. He directed the forensic serology laboratory and performed and directed forensic biological analyses in all New York City homicide investigations. In the wake of the World Trade Center (WTC) attacks on September 11, 2001, he assumed the responsibility for identifying the people who perished. He designed, organized, and implemented the DNA testing strategy that became the cornerstone for the majority of the identified victims. When the New York City Office of the Chief Medical Examiner effort to identify the WTC victims paused, he accepted a professorship in the Biochemistry and Molecular Biology Department and the directorship of the forensic science program at the Pennsylvania State University.

Elizabeth A. Thompson is a professor in the Department of Statistics and adjunct professor in the departments of Biostatistics and of Genome Sciences at the University of Washington, and Director of an Interdisciplinary Graduate Certificate program in Statistical Genetics. She received her B.A. in mathematics and Ph.D. in mathematical statistics from Cambridge University, UK, and did postdoctoral work in the Department of Genetics, Stanford University, before taking up a position on the faculty of the Department of Pure Mathematics and Mathematical Statistics at the University of Cambridge in 1976. She joined the faculty of the University of Washington in December 1985 as a professor of statistics and served as chair 1989-1994. Dr. Thompson's research is in the development of methods for model-based likelihood inference from genetic data, particularly from data observed on large and complex pedigree structures both of humans and of other species, and including inference of relationships among individuals and among populations. Dr. Thompson is a recipient of a Doctor of Science degree from the University of Cambridge, the Jerome Sacks award for cross-disciplinary research from the National Institute for Statistical Science, the Weldon Prize for contributions to Biometric Science from Oxford University, UK, and a Guggenheim fellowship. She has served on the NRC Committee on Applied and Theoretical Statistics and on the Scientific Advisory Boards of the Pacific Institute for Mathematical Sciences, the Banff International Research Station, and the Institute for Pure and Applied Mathematics. She also serves on several committees of the International Biometric Society, including as a member of Council. Dr. Thompson is an elected member of the International Statistical Institute, the American Academy of Arts and Sciences, and the National Academy of Sciences.

Kasthuri Venkateswaran is a senior research scientist at the California Institute of Technology's Jet Propulsion Laboratory. His 32 years of research encompass marine, food, and environmental microbiology. He has applied his research in molecular microbial analysis to better understand the ecological aspects of microbes, while conducting field studies in several extreme environments such as deep sea (2,500 m), pristine caves (3,000 m altitude), spacecraft (Mars Odyssey, Genesis, MER, Mars Express, Phoenix, MSL) assembly facility clean rooms (various NASA and European Space Agency facilities), as well as the space environment in Earth orbit (International Space Station). Of particular interest are microbe-environment interactions with emphasis on the environmental limits in which organisms can live. The results are used to model microbe-environment interactions with respect to microbial detection and the technologies to rapidly monitor them without cultivation. The bioinformatics databases generated by Dr. Venkateswaran are extremely useful in the development of biosensors. Further, these models or information in databases are extrapolated to what is known about the spacecraft surfaces and enclosed habitats in an attempt to determine forward contamination as well as develop countermeasures (develop cleaning and sterilization technologies) to control problematic microbial species. Specifically, his research into the analysis of clean room environments using state-of-the-art molecular analysis coupled with nucleic acid and protein-based microarrays will enable accurate interpretation of data and implementation of planetary protection policies of present missions, helping to set standards for future life-detection missions.

David R. Walt is Robinson Professor of Chemistry and Professor of Biomedical Engineering at Tufts University and is a Howard Hughes Medical Institute Professor. He received a B.S. in Chemistry from the University of Michigan and a Ph.D. in Chemical Biology from SUNY at Stony Brook. His laboratory applies micro- and nanotechnology to urgent biological problems such as the analysis of genetic variation and the behavior of single cells, single molecule detection, as well as the practical application of arrays to the detection of explosives, chemical and biological warfare agents, and food and waterborne pathogens. Dr. Walt is the Scientific Founder and a Director of both Illumina Inc. and Quanterix Corp. He has received numerous national and international awards and honors for his fundamental and applied work in the field of optical sensors and arrays. He is a member of the National Academy of Engineering, a fellow of the American Institute for Medical and Biological Engineering and the AAAS. He has served on a number of NRC committees including the Committee on Review and Evaluation Methodology for Biological Point Detectors.

Staff

Anne-Marie Mazza is the Director of the Committee on Science, Technology, and Law. Dr. Mazza joined the National Research Council in 1995. She has served as Senior Program Officer with both the Committee on Science, Engineering, and Public Policy and the Government-University-Industry Research Roundtable. In 1999 she was named the first director of the Committee on Science, Technology, and Law, a newly created activity designed to foster communication and analysis among scientists, engineers, and members of the legal community. Dr. Mazza has been the study director on numerous Academy reports including *Managing University Intellectual Property in the Public Interest* (2010); *Strengthening Forensic Science in the United States: A Path Forward* (2009); *Science and Security in a Post- 9/11 World* (2007); *Daubert Standards: Summary of Meetings* (2006); *Reaping the Benefits of Genomic and Proteomic Research: Intellectual Property Rights, Innovation, and Public Health* (2005); *Intentional Human Dosing Studies for EPA Regulatory Purposes: Scientific and Ethical Issues* (2004); *Ensuring the Quality of Data Disseminated by the Federal Government* (2003). Dr. Mazza received an NRC distinguished service award in 2008. In 1999-2000, Dr. Mazza divided her time between the National Academies and the White House Office of Science and Technology Policy (OSTP), where she served as a Senior Policy Analyst responsible for issues associated with a Presidential Review Directive on the government-university research partnership. Before joining the Academy, Dr. Mazza was a Senior Consultant with Resource Planning Corporation. Dr. Mazza received a B.A., M.A., and Ph.D. from the George Washington University.

Frances E. Sharples has served as the Director of the National Research Council's Board on Life Sciences since October 2000. Immediately prior to this position, she was a Senior Policy Analyst for the Environment Division of the White House Office of Science and Technology Policy (OSTP) for four years. Dr. Sharples came to OSTP from the Oak Ridge National Laboratory, where she served in various positions in the Environmental Sciences Division between 1978 and 1996, most recently as a Research and Development Section Head. Dr. Sharples received her B.A. in Biology from Barnard College and her M.A. and Ph.D. in Zoology from the University of California, Davis. She served as an AAAS Environmental Science and Engineering Fellow at EPA during the summer of 1981 and as a AAAS Congressional Science and Engineering Fellow in the office of Senator Al Gore in 1984-85. She was a member of NIH's Recombinant DNA Advisory Committee in the mid-1980s and was elected a Fellow of the AAAS in 1992.

Ericka D. Martin McGowan is program officer with the National Research Council Board on Chemical Sciences and Technology (BCST), where she contributes to scientific policy studies related to the detection of biological and chemical warfare agents as well as issues at the interface of chemistry and biology. Since joining the NRC in 2004, Mrs. McGowan has been involved with the following NRC studies and reports: *BioWatch and Public Health Surveillance: Evaluating Systems for the Early Detection of Biological Threats* (2010); *Test and Evaluation of Biological Standoff Detection Systems* (2008); *Protecting Building Occupants and Operations form Biological and Chemical Airborne Threats: A Framework for Decision Making* (2007); *Exploring Opportunities in Green Chemistry and Engineering Education: A Workshop Summary to the Chemical Sciences Roundtable* (2007); *Measuring Respirator Use in the Workplace* (2007) *Terrorism and the Chemical Infrastructure: Protecting People and Reducing Vulnerabilities* (2006). Mrs. McGowan received a B.S. in Biology, minor in Chemistry, from Southern University and A&M College, and an M.S. in Public Health Microbiology and Emerging Infectious Diseases from the George Washington University.

Steven Kendall is Associate Program Officer for the Committee on Science, Technology, and Law. He is a Ph.D. candidate in the Department of the History of Art and Architecture at the University of California, Santa Barbara, where he is completing a dissertation on 19th century British painting. Mr. Kendall received his M.A. in Victorian Art and Architecture at the University of London. Prior to joining the NRC in 2007, he worked at the Smithsonian American Art Museum and the Huntington in San Marino, California.

Amanda Cline is an administrative assistant on the Board on Chemical Sciences and Technology. She joined the NRC in 2007, after receiving her B.S. in environmental studies from Bucknell University. Ms. Cline has worked in the report review office of the Division on Earth and Life Studies and as a program assistant for the Board on Life Sciences, where she supported the Human Embryonic Stem Cell Advisory Committee, the Interstate Alliance on Stem Cell Research, the Committee on a New Biology for the 21st Century, the Committee on Ecological Impacts of Climate Change, and other NRC activities.

Kathi E. Hanna has over 25 years of experience in science, health, and education policy as an analyst, writer, and editor. In the 1990s Dr. Hanna served as Research Director and Editorial Consultant to President Clinton's National Bioethics Advisory Commission. She also served as Senior Advisor on Reproductive Toxicology to the President's Advisory Committee on Gulf War Veterans Illnesses. She was the lead analyst and author for President Bush's Task Force to Improve Health Care Delivery for Our Nation's Veterans and served in a similar capacity for the Task Force on the Future of Military Health Care. In the 1980s and 1990s, Dr. Hanna was a Senior Analyst at the congressional Office of Technology Assessment, contributing to numerous science policy studies requested by congressional committees on science education, research funding, science and economic development, biotechnology, women's health, mental health, children's health, human genetics, bioethics, cancer biology, and reproductive technologies. In the past two decades she has served as an analyst and editorial consultant to the Howard Hughes Medical Institute, the NIH, the NRC, the U.S. Office for Human Research Protections, FasterCures, the Lance Armstrong Foundation, the American Heart Association, the Burroughs Wellcome Fund, the March of Dimes, the U.S. Anti-Doping Agency, and biotechnology companies. She has been a consultant and lead author on NIH strategic plans; ad hoc committees of the Advisory Committee to the Director, NIH; the Secretary's Advisory Committee on Genetics, Health, and Society; and the NIH Office of Behavioral and Social Science Research. She has authored or coauthored over 40 reports of studies of children and environmental health, obesity, immunization, genetics, emergency care, epilepsy, cancer, forensic science, and general health and science policy. Before moving to the Washington, D.C., area, she was the Genetics Coordinator/Counselor at Children's Memorial Hospital in Chicago, where she directed clinical counseling in pediatric genetics and coordinated an international research program in prenatal diagnosis. Dr. Hanna received an A.B. in biology from Lafayette College, an M.S. in human genetics from Sarah Lawrence College, and a Ph.D. in government and health services administration from the School of Business and Public Management at George Washington University.

Cameron H. Fletcher is the Managing Editor of the *ILAR Journal*, the quarterly, peer-reviewed official publication of the National Research Council's Institute for Laboratory Animal Research. In her 25 years at the NRC she has edited numerous reports for the Division on Earth and Life Studies, Division on Behavioral and Social Sciences and Education, Division on Engineering and Physical Sciences, and Policy and Global Affairs. She has also edited reports and other publications for the Pew National Commission on Industrial Farm Animal Production, the Peterson Institute for International Economics, the Congressional Budget Office, the International Association of Oil and Gas Producers, and Columbia University. Before her tenure at the NRC she taught French, Spanish, Latin, and English at private schools in Connecticut and Rhode Island. She received her AB *cum laude* from Bryn Mawr College.

Index

A

agar, 89, 113, 167
American Media, Inc. (AMI), 26, 56–57, 60, 62, 65, 67, 76
American Society of Microbiology (ASM) Bio Defense Meeting Presentations, 178
Amerithrax investigations, 25–26
    analytical techniques used on evidentiary material, 57, 58–59t
    collection and analysis of clinical and environmental samples and cross contamination, 60, 176–77
    clinical and epidemiological samples, 60–62, 64
    crime scene environmental samples, 64–66
    letter material and cross contamination, 67–70
    samples from overseas site identified by intelligence, 66–67
    federal coordinated response and assignment of laboratory work, 55–57
    See also specific topics
Ames Ancestor sequence, 103
Ames strain *B. anthracis*, 6–7, 31, 32, 44, 103, 129, 169
Ames strain DNA, 8
Ames strain identification, FBI documents regarding, 162–63
Ames strain samples, subpoena protocol for collection and submission of, 126–30, 132, 144–47
amplified fragment length polymorphism (AFLP), 98

anthrax mailings, 40–41
    size and granularity of material in letters, 79–80
    trajectory and outcomes of, 61–62, 62f
anthrax mailings case
    background, 25–26
    chronology of, 26, 30–31
        timeline of key events, 28–30t
        timeline of scientific events in, 48–52t
anthrax program, 66
Armed Forces Institute of Pathology (AFIP), 56, 57, 66, 79, 81–84, 95

B

Bacillus sp.
    B. anthracis, 1, 37, 44–45, 97
        as biological weapon, 40–41
        biology, 37–38, 44–45
        chemical and physical properties, 177
        clinical aspects, 38–40
        early history of Ames strain of, 44
        identification of *B. anthracis* strain, 97–100
        isolates (*see* morphotype isolates)
        modes of transmission, 39
        phylogeny, 41–44
        worldwide distribution of lineages of, 43, 43f
    B. cereus, 42, 84, 88
    B. subtilis, 84, 96, 121–22, 169–70
        contamination of New York samples with, 65, 104–6
        genetic diversity and phylogenetic characterization of, 171

in *New York Post* letter, 96, 105, 121–22
whole genome assembly of *B. subtilis* isolate, 170–71
B. subtilis screening, 171–72
bacterial growth conditions and processing methods, features of, 87–89
bioterrorism investigations, 53, 54
Blanco, Ernesto, 26
blood agar, 89, 113, 167
Brokaw, Tom, 26

C

carbon-14 (^{14}C) dating, 165–66. *See also* radiocarbon dating
Center for Accelerator Mass Spectrometry (CAMS), 90
Centers for Disease Control and Prevention (CDC), 126
Chemical, Biological, Radiological, and Nuclear (CBRN) Sciences Unit/Chemical Biological Science Unit (CBSU), 71
chemical analysis
   committee findings, 93–96
   methods for, 81, 81t
   See also specific topics
Chemistry Unit, FBI Laboratory, 165
Committee on Review of the Scientific Approaches Used During the FBI's Investigation of the 2001 *Bacillus anthracis* Mailings
   biographical information on, 193–204
   charge to, 27
   committee process, 33–35
   findings, 4–23t, 70–74, 121–23
   formation, xi
   recommendations, 10, 70–74
   See also specific topics
cross contamination. *See under* Amerithrax investigations

D

Daschle, Tom, letter received by, 26, 28t, 30–31, 58–59t, 68, 79–80, 82, 87, 113–22. *See also specific topics*
"deep sequencing," 101, 150
Defence Research Establishment Suffield (DRES), 69
Department of Justice (DOJ)
   Amerithrax Investigative Summary, 93, 147
   scientific conclusions, 11–23t, 32–33
diatrizoate, 38
detection of, 77, 87–89, 96
dilution experiments, 188–89
DNA, 8, 102
Dugway Proving Ground (DPG), 78, 95–96, 168–69

E

edema factor (EF), 39
Ekaterinburg. *See* Sverdlovsk outbreak
envelope measurements, 92–93
enzymes produced by *B. anthracis*

F

FBI (Federal Bureau of Investigation), 9
   documents provided by, 161–79
   declassified reports, 177
   scientific conclusions and committee findings, 4–23t, 32–33
   scientific investigations, 31–32 (*see also* Amerithrax investigations)
   See also specific topics
FBI hazardous materials (HAZMAT) teams, 64, 68
FBI Repository (FBIR), 32
   creation of, 126–30
FBI Repository (FBIR) samples
   comparison of material in letters with, 125–26
      analyses based on resampling RMR-1029 and interpretation of results, 140–44
      committee findings, 144–51
      See also specific topics
   statistical interpretation of the evidence and analyses of, 132–34

G

genetic engineering, 100, 102–4, 163
genetic markers in *New York Post* letter (powder), 115t, 116–22, 148–49

genome assembly of *B. subtilis* isolate, 170–71
genome sequencing, 101, 163–64
   of morphotype isolates, 114–19
   See also Institute of Genomic Research
genotypes, 139, 139t
   A1 and A3, 119, 172–73
   B and D, 119–20, 173–75
   development and application of assays for, 119–21
   E, 120–21, 175–76
   genetic assays to test for the four, 130
   mutation, in FBIR samples, 133–34, 133t, 134t
   observed and expected distribution of positive signatures for the four, 137, 138t
   in RMR-1029, 125, 130–32, 138, 139t, 140–42, 145–48

H

hazardous materials (HAZMAT) teams, 64, 68
heme, 89, 167

I

inductively coupled plasma-optical emission spectroscopy (ICP-OES), 81–83, 94
inhalational anthrax, 26, 28–29t, 30, 31, 39, 40, 44–45, 60–62, 64, 97
Institute of Genomic Research (TIGR), 32, 102–5, 115, 117–20
Institute of Infectious Diseases. *See* U.S. Army Research Institute of Infectious Diseases
Ivins, Bruce, 26, 140–42, 145

J

Justice Department, U.S. *See* Department of Justice

K

Keim, Paul, 99–100

L

Lawrence Livermore National Laboratory (LLNL), 79, 86, 90
Leahy, Patrick, letter received by, 30, 68, 69, 76–80, 88–92, 96, 109, 113–22
   powder on, 63f
   silicon content, 82–84, 85f, 87, 94, 95
lethal factor (LF), 39
letter material, silicon and other elements in, 80
   elemental analysis, 81–84
letter powder
   Leahy, 62, 63f, 64
   New York Post, 62, 63f, 64
Los Alamos National Laboratory (LANL), 100

M

mass spectrometry (MS), 88–90
media component analysis, 89
meglumine, 38
   detection of, 77, 87–89, 96
morphological variants in evidentiary material, identification and characterization of
   committee findings, 121–23
   development and application of assays for genotypes, 119–21
   selection criteria for genetic variations used in screening, 113–14
   See also morphotypes
morphotype isolates, whole genome sequencing of, 114–19
morphotypes, 5–6
   background information on, 107–9
   defined, 106
   detection and characterization of, 109–13
   phenotypic characteristics, 113, 113t
   genetic characterization of, 116, 116t
   reasons FBI was interested in, 106–7
multiple-locus VNTR analysis (MLVA), 98, 99

N

nano time-of-flight secondary ion mass spectrometry (nano-SIMS), 86
National Academies, xii–xiii, 35
National Academy of Sciences (NAS), xi, 56
National Research Council (NRC), xi, 1, 26
New York City letters, 26, 60–62. *See also* American Media, Inc.

New York Post letter (powder), 62, 63f, 64, 68, 85f, 94–95
  B. subtilis in, 96, 105, 121–22
  genetic markers in, 115t, 116–22, 148–49
  SEM-EDX analysis of, 83–85, 94–96

P

plasmids, 39
polymerase chain reaction (PCR) technique, 102
polymorphism(s)
  amplified fragment length, 98
  single nucleotide, 115–18
postal workers, 61
protective antigen (PA), 39

R

radiocarbon dating, 181–82
  of *B. anthracis* samples, 90
  of letter received by Patrick Leahy, 95–96
  See also carbon-14 (^{14}C) dating
RenoCal, 88, 168
RMR-1029 (spore-containing flask), 32, 74, 77, 85, 88, 96, 149, 150
  analyses based on resampling, 140–44
  genotypes in, 125, 130–32, 138, 139t, 140–42, 145–48
  results obtained by resampling from, 142, 143t
RMR-1029 spores, derivation of, 130–32
RMR-1030 (spore-containing flask), 85n

S

Sandia National Laboratories (SNL), 83, 84, 164–65
scanning electron microscope. *See* SEM
science
  FBI's uses of, 35–36
  qualifiers of certainty in biological sciences, 53, 55
  and scientific investigation, as part of law enforcement investigation, 47, 53–55
"Select Agents" program, 126
SEM (scanning electron microscope), 79, 81, 85f

SEM-EDX (scanning electron microscope with energy-dispersive X-ray) analysis, 79, 81–86
  of Leahy powder, 85f
  of *New York Post* letter, 83–85, 94–96
Senate letters, 26, 30–31
silicon analysis, 7–8, 12t, 84–87, 94–96. *See also under* letter material
silicon measurements in evidentiary and surrogate samples, 82, 82t
single nucleotide polymorphisms (SNPs), 115–18
Soviet Union. *See* Sverdlovsk outbreak
spatially resolved elemental analysis, 83–84
spo0A gene, 108–9
spo0F gene, 117
spore preparation
  estimates of media volume required for, 77, 77t
  and purification, 75–78
  time needed for, 8
spores
  biology, 37–38, 44–45
  derivation of RMR-1029, 130–32
  estimated ranges of total number of, 76, 76t
  resilience, 37–38
stable isotope analysis, 90–93, 166–67
  forensics potential, 183–84
Stable Isotope Ratio Facility for Environmental Research (SIRFER), 90–93
Statistical Analysis Report (FBI), 135–36
  committee assessment of, 185–91
  representativeness, randomness, and independence, 136–40, 185–86
Stevens, Robert, 26, 28t, 60
subpoena protocol for collection and submission of Ames strain samples, 126–30, 132, 144–47
surrogate preparation and purification, 78–79
Sverdlovsk outbreak, 41

T

TaqMan technique, 105, 106
Technical Review Panels, 56

U

U.S. Army Research Institute of Infectious Diseases (USAMRIID), 56–57, 66, 109, 131, 140–41, 161–62
USDOJ. *See* Department of Justice

V

variable number tandem repeat (VNTR) analysis, 98, 99
volatile organic chemicals (VOCs), 89–90

W

water samples, stable isotope analysis of, 92